The Institute of Biology's
Studies in Biology no. 166

Plants and Nitrogen

Owen A. M. Lewis

MSc, PhD, FRSSAf, FIBiol, FLS

Professor of Botany
University of Cape Town

R

Edward Arnold

© Owen A. M. Lewis 1986

First published in Great Britain 1986 by
Edward Arnold (Publishers) Ltd, 41 Bedford Square, London WC1B 3DQ

Edward Arnold (Australia) Pty Ltd, 80 Waverley Road, Caulfield East,
Victoria 3145, Australia

Edward Arnold, 3 East Read Street, Baltimore, Maryland 21201, USA.

British Library Cataloguing in Publication Data

Lewis, Owen A.M.
    Plants and nitrogen.——(The Institute of
Biology's studies in biology, ISSN 0537-9024; no. 166)
    1. Nitrogen——Metabolism   2. Plants—Metabolism
    I. Title   II. Series
    581.1'33   QK898.N6
    ISBN 0-7131-2899-2

Text set in 9½/11 pt English Times Compugraphic
by Colset Pte Ltd, Singapore

Printed and bound in Great Britain at
The Camelot Press Ltd, Southampton

General Preface to the Series

Because it is no longer possible for one textbook to cover the whole field of biology while remaining sufficiently up to date, the Institute of Biology proposed this series so that teachers and students can learn about significant developments. The enthusiastic acceptance of 'Studies in Biology' shows that the books are providing authoritative views of biological topics.

The features of the series include the attention given to methods, the selected list of books for further reading and, wherever possible, suggestions for practical work.

Readers' comments will be welcomed by the Institute.

1986   Institute of Biology
                                        20 Queensberry Place
                                        London SW7 2DZ

Preface

Nitrogen in its various combined forms is a precious commodity in the biosphere where it forms a major component of so many key compounds. In nearly all regions of the world where water supply or sunlight do not limit growth, it is the availability of inorganic nitrogen in the soil that determines biological productivity. Thus the study of nitrogen in the biosphere is not only a fascinating one for life scientists but an economically and ecologically rewarding one, too.

This book attempts to survey in a simple fashion the processing of nitrogen in the biosphere, particularly its plant component, by following the steps of the nitrogen cycle, that major global process which conserves our planet's limited supply of combined nitrogen and mediates its oscillation between animate and inanimate matter. In so doing it is hoped to hightlight not only the biochemical and physiological subtleties of the processes involved but also their ecological and agricultural significance.

Cape Town 1986   O.A.M.L.

Contents

1 Nitrogen in the Biosphere

Without nitrogen, life on this planet as we know it today could not exist. This is because nitrogen is a major component of a number of compounds which are essential for the structure and functioning of biological organisms, from the primitive prokaryotes to the highly evolved and sophisticated eukaryotes. To understand the intense interest of the biologist in this remarkable element one need only appreciate the vital importance of the nucleic acids in perpetuating the characteristics of an organism and directing its growth, development and metabolism; the pivotal position of proteins (as enzymes) in executing the instructions of the nucleic acids, or building the mobile framework of the protoplasm and other essential structures of the organism; and the fundamental roles of compounds such as adenosine triphosphate (ATP) in energy transfers, or nicotinamide adenine dinucleotide (NAD) and the flavoproteins in the redox reactions so crucial in biological metabolism.

It is a curious fact that, apart from water, nitrogen is the key substance that limits the primary productivity of the plant kingdom in most parts of the world. We know that nitrogen is a constituent of many important plant compounds, but we also know that nitrogen is one of the commonest elements on Earth. Why then, should it occupy this growth-limiting role in the biosphere?

The answer to this enigma lies in the inert dinitrogen (N_2) molecule, which most biological systems are unable to utilize in their metabolism. The stability of the $N \equiv N$ combination in its high oxidation state ($+5$) is too great for the normal assimilation processes of the plant to upset in order to synthesize its own nitrogenous compounds (hence Lavoisier's French name for nitrogen, 'azote', meaning 'without life'). Thus, the plant has to rely on nitrogen occurring in more reactive combinations, e.g. with hydrogen (as in ammonia) or oxygen (as in nitrate or nitrite), to build its organic nitrogen-containing molecules. Such nitrogen is referred to as 'combined nitrogen' and unfortunately is not nearly as common as the dinitrogen state. In fact, geochemists have calculated that of earth's total supply of nitrogen (some 57.4×10^{18} kg), only 0.0025 per cent (1.5×10^{15} kg) is available in the biosphere for plant growth. The vast bulk (93%) is locked away in the rocks of the earth's mantle, while 7 per cent (3.8×10^{18} kg) is found as dinitrogen in the earth's atmosphere. Of the 0.0025 per cent combined nitrogen present and available for biological use, 57 per cent occurs in inorganic molecules while 43 per cent occurs in the organic form.

Biologically available nitrogen, therefore, forms only a minute fraction of the total present on earth. Furthermore, much of this nitrogen is temporarily inaccessible to plants as it is present in the form of dead organic material from which it is only slowly released by microbial action. Figure

1-1 shows the quantitative relationship between the distribution of nitrogen in dead (necromass) and living organic material (biomass) in the terrestrial biosphere. Details for the marine biosphere, where over 60 per cent of total organic nitrogen is found, are unfortunately not yet available.

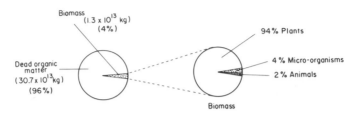

Fig. 1-1   Distribution of nitrogen in dead and living organic material in the terrestrial biosphere. The left hand circle shows the quantitative relationship between the nitrogen of the biomass (4%) compared with the nitrogen of the necromass (96%). The right hand circle shows the distribution of nitrogen between the plants (94%), animals (2%) and micro-organisms (4%) which make up the biomass.

Thus, we find that in the biosphere there is a real shortage of the nitrogen form that can be made use of by living organisms and a real necessity exists for the conservation of this material. Nature responds to this necessity by recycling combined nitrogen in a complex process known as the 'Nitrogen Cycle', a simplified version of which is shown in Fig. 2-1.

2 The Cycling of Nitrogen in Nature

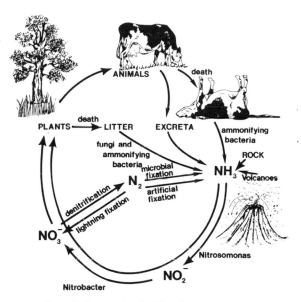

Fig. 2-1 The recycling of nitrogen in the biosphere.

Basically, the nitrogen cycle involves the scavenging of combined nitrogen from dead plants and animals by microbial action, its passage through the soil in various forms and its final re-absorption by plants to form the proteins, nucleic acids and other nitrogenous compounds of the living cells of plants and animals. In this cycle combined nitrogen can be lost to the major nitrogen pool of the biosphere, the earth's atmosphere, by a process called denitrification and gained from it by natural and artificial nitrogen fixation. The major processes of the nitrogen cycle in which various soil and symbiotic microbes play a dominant role are discussed below.

2.1 Mineralization of Nitrogen

When plants and animals die, their nitrogenous remains become *mineralized*; i.e. converted into inorganic forms of nitrogen, mainly ammonia, by a host of bacteria, fungi and actinomycetes. In the degradation of nitrogen-rich material (e.g. animal excreta, flesh) anaerobic bacteria are often involved, resulting in the production of foul-smelling products such as amines, mercaptans and hydrogen sulphide as well as ammonia. This process is known as *putrefaction*. Aerobic degradation results in the production of less obnoxious end-products, particularly ammonia.

Typical mineralizing organisms are summarized in the following table.

Ammonifying bacteria	Actinomycetes	Fungi
Pseudomonas spp	*Streptomyces* spp.	*Alternaria* spp.
Bacillus spp.		*Aspergillus* spp.
Clostridium spp.		*Mucor* spp.
Serratia spp.		*Penicillium* spp.
Micrococcus spp.		*Rhizopus* sp.

Many ammonifying organisms release protein-hydrolysing enzymes (proteases) when invading decomposing organic material. This results in a rapid hydrolysis of the protein into amino acids which can be absorbed by the invading organisms and further hydrolysed inside their cells. Some microorganisms do not, however, secrete enzymes and rely on other organisms to do the initial protein breakdown for them. Many microorganisms also secrete nucleic acid hydrolysing enzymes (ribonucleases and deoxyribonucleases) which degrade deoxyribonucleic acid (DNA) and ribonucleic acid (RNA), the nitrogen of which ends up as ammonia.

The nitrogenous compounds of plants, particularly those which form complexes with materials such as polyphenols, may take months or years to mineralize (for example, many forms of plant litter). This slowly mineralizing material forms the humus content of soils.

2.1.1   Measurement of the rate of net mineralization

This can be achieved by taking a soil, freeing it of higher plants, measuring its inorganic nitrogen (NH_4 + NO_3) and incubating it for several days. The inorganic nitrogen is then redetermined and the difference represents the rate of mineralization of that soil.

Net mineralization always represents the difference between the *total* mineralization and the immobilization activities of the soil microflora. In any particular soil, a change in inorganic nitrogen can be represented by the following formula.

$$N_i = \text{organic N mineralized} - (N_a + N_p + N_l + N_a)$$

where
N_i = inorganic N
N_a = N assimilated by microflora
N_p = N removed by higher plants
N_l = N lost by leaching
N_d = N volatilized

2.1.2   Factors affecting the rate of mineralization

There are three major environmental factors which can affect the rate at which the nitrogenous contents of plant and animal remains are converted into the inorganic form. These are as follows.

(*a*) *The water content of soil*   Mineralization is very slow in dry soils, but improves as water content rises (see Fig. 2-2). In climatic zones with wet-dry cycles, the onset of rainfall is associated with a rapid increase in the rate of mineralization. For some as yet unknown reason, considerable quantities of organic nitrogen can be mineralized in a sequence of dry and wet cycles – far greater than if the soils were kept permanently moist.

(*b*) *The aeration of the soil*   The effect of soil aeration depends on whether the mineralizing organisms are aerobic or anaerobic. Most soils do show a falling off in mineralization rate when the moisture content of the soil is above 70 per cent of the water-holding capacity, due to poor aeration (see Fig. 2-2). But in wet paddy fields, mineralization is very rapid due to the presence of anaerobic mineralizing organisms.

(*c*) *Temperature*   Most of the organisms involved in mineralization appear to be thermophilic rather than mesophilic and the optimum temperature for ammonification is usually between 40°C and 60°C. Ammonium accumulates rapidly in compost and manure heaps which often maintain themselves at 65°C due to the heat given off by respiration. Freezing followed by thawing increases the degradation rate of humus. The reason for this is again not properly understood.

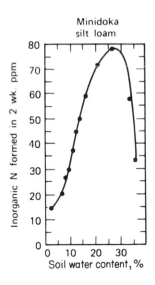

Fig. 2-2   Effect of soil water content on inorganic nitrogen accumulation (N mineralization) in Minidoka silt loam. (From Alexander, M. (1977). *Soil Microbiology*, 2nd ed. John Wiley & Sons, New York.)

2.1.3   Volatilization of Ammonia

This occurs if mineralization of nitrogen-rich materials takes place rapidly. The ammonium is produced in such abundance that the pH of the soil rises into the alkaline range converting ammonium (a soluble ion) into ammonia (a gas) which is then lost to the soil. Loss of ammonium by volatilization is also a problem in naturally alkaline soils, for example, the chalk soils of Europe.

2.2   Microbial immobilization of nitrogen

The final product of mineralization, ammonia, does not necessarily become immediately available for the nutrition of the higher plant. It is often immobilized by the soil microflora which use it for their own growth and multiplication. The degree to which the ammonia released by mineralization can become immobilized in this way depends largely on the carbon:nitrogen (C:N) ratio of the decomposing material. If this material is rich in carbohydrate, so much carbon is available for microbial respiration and to provide skeletons for organic molecules that all the mineralized N is used up in producing microbial protoplasm. In fact, the inorganic N present originally in the soil may also be used up, resulting in a *depletion* of the soil's inorganic nitrogen available for plant growth. (This effect may be demonstrated by simply adding glucose to a soil as shown in Fig. 2-3.) It may be many weeks before the newly arrived nitrogen becomes available

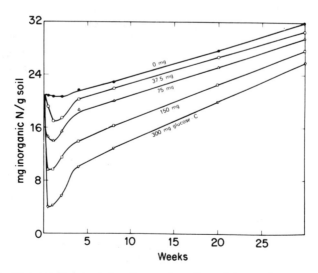

Fig. 2-3   Changes in inorganic nitrogen in soil receiving various quantities of glucose (From Alexander, M. (1977). *Soil Microbiology*, 2nd ed. John Wiley & Sons, New York.)

Table 2-1 Nitrate level of soils incubated with materials of varying nitrogen contents. (Lyon, T.L., Bizzell, J.A. and Wilson, B.D. (1929). *J. Amer. Soc. Agron.*, **15**, 457–67.)

Organic amendment	Nitrogen content of substrate[a]	After three months Nitrate-N in soil	N Change
	%	mg	mg
Untreated soil	–	947	–
Dried blood	10.71	1751	+ 804
Clover roots	1.71	924	– 23
Corn roots	0.79	511	– 436
Timothy roots	0.62	398	549
Oat roots	0.45	207	– 740

[a] All materials applied at rates to give 600 mg N.

for plant growth. This follows mineralization of the microbial population when it dies off after the exhaustion of the carbohydrate supply.

If, however, the C:N ratio of the decomposing material is high, then excess ammonium is available to enter the inorganic N fraction of the soil from which it may be absorbed by higher plants. The effect of adding materials of various C:N ratios is shown in Table 2-1.

Experimentation has shown that plant residues which contain more than 1.8 per cent N (this represents a C:N ratio of below 20:1) usually enrich a soil immediately with inorganic nitrogen; those with a nitrogen content of 1.8 per cent to 1.2 per cent *deplete* a soil temporarily, while those with an even lower N content may deplete the soil's inorganic nitrogen for a prolonged period of time.

2.3 Nitrification

In many soils ammonium, produced by mineralization, does not accumulate because of the activity of the *nitrifying bacteria* which rapidly convert ammonium into nitrite and then into nitrate. These bacteria are *chemotrophs* and are completely dependent on this oxidation process for their energy supply.

There are two genera of bacteria which have been shown to be important in this process (although others, e.g. *Nitrosolobus* and *Nitrocystis*, do exist).

(a) *Nitrosomonas* which converts ammonia to nitrite (one species, *N. europaea*, forming ellipsoidal or short rods)

$$NH_3 + \tfrac{1}{2}O_2 \quad \rightarrow NH_2OH \qquad\qquad +15 \;\; kJ$$

$$NH_2OH + O_2 \quad \rightarrow NO_2 + H_2O + H^+ \;\; -289 \;\; kJ$$

(b) *Nitrobacter* which converts nitrite into nitrate (one species, *N. winogradskyi*, a short rod bacterium)

$$NO_2^- + \tfrac{1}{2}O_2 \quad \rightarrow NO_3 \qquad\qquad\qquad -77 \;\; kJ$$

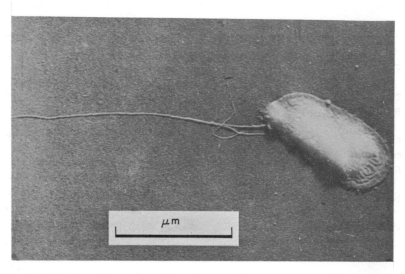

Fig. 2-4   Electron photomicrograph of *Nitrosomonas europea* (From Alexander, M. (1977). *Soil Microbiology*, 2nd ed. John Wiley & Sons, New York.)

Most crop plants usually receive nitrogen in the form of nitrate, as the fertilizer applied is either in this form or, if in the ammonia or urea form, is rapidly oxidized to nitrate.

The advantage of nitrate to the plant is that it is non-toxic (ammonium is, except at low concentrations). The disadvantage is that it is a negatively charged ion and consequently it is easily leached from the soil (which is also negatively charged) by rain and washed into rivers which it then pollutes. Also, the higher plant has to use large quantities of energy in reducing it back to ammonia for use by the plant. In some parts of the world, chemicals (e.g. nitrapyrin) are added with fertilizer to retard the process of nitrification by poisoning *Nitrosomonas*. Nitrate and ammonia as nitrogen sources for plant growth are further discussed in Chapter 3.

2.3.1   Factors affecting nitrification

The main environmental factors that influence the process of nitrification are:

(*a*) *pH*   In acid environments nitrification proceeds very slowly and the responsible bacteria are rare or totally absent. The response to acidity appears to depend on the strains of bacteria involved. Some soils show a marked fall-off in nitrification below pH 6.0 with none at all below pH 5.0, whereas some soils may exhibit a degree of nitrification even below pH 4.0. The acidity affects not only the process itself but also the microbial numbers. Nitrification in acid soils is markedly enhanced by liming. Thus we find that in acid soils, ammonium is usually the predominant form of inorganic N, whereas in neutral or alkaline soils, nitrate is the dominant form. In many grasslands of the world (e.g. the Zimbabwian high-veld) ammonium is present in greater abundance than nitrate. In addition to a low soil pH in these regions it has also been suggested that the roots of grasses secrete compounds which suppress nitrification.

(*b*) *Aeration*   Oxygen is absolutely essential for nitrification and in poorly aerated soils (e.g. waterlogged soils) nitrification is heavily suppressed.

(*c*) *Soil moisture*   Excess water as found in waterlogged soils suppresses nitrification because of lack of oxygen. In dry soils, however, there is not enough moisture for bacterial metabolism and the moistening of such soils rapidly increases the biosynthesis of nitrate.

(*d*) *Temperature*   Nitrification is markedly affected by temperature. Below 5°C and above 40°C the rate is very slow with an optimum temperature between 30° and 35°C. So in winter the rate of nitrification is much lower than in summer.

In many soils, then, the progress of combined nitrogen through the nitrogen cycle is as follows.

(*i*) Mineralization of dead animal/plant remains by ammonifying bacteria.

(*ii*) Immobilization of nitrogen by soil bacteria (depending on the C:N ratio of the decomposing material).

(*iii*) Remineralization of bacterial nitrogen to ammonia on death of soil bacteria.

(*iv*) Nitrification of ammonia to nitrate by nitrifying bacteria (depending largely on soil pH).

(*v*) Absorption of nitrate or ammonia by the plant for assimilation into living material.

2.4   Losses of combined nitrogen from the nitrogen cycle

The major loss of combined nitrogen to the atmospheric dinitrogen pool is by the process of denitrification. Strangely enough, there are no micro-organisms that are *specifically* engaged in carrying out this process. Instead, we find that a number of ammonifying bacteria which under normal aerobic conditions use O_2 as their electron acceptor in respiration, switch to using nitrate when O_2 is in short supply, for example, when the soil becomes waterlogged. The nitrate becomes reduced to either N_2 gas, or the gaseous oxides of N, nitrous oxide (N_2O), nitric oxide (NO) or nitrogen dioxide (NO_2), which are volatilized and lost to the soil. The commonest denitrifiers are, *Pseudomonas* spp., *Paracoccus* spp. and *Thiobacillus denitrificans*.

Fungi are not often associated with denitrification. Denitrification is an extremely serious form of combined nitrogen loss from soils, thousands of tonnes of N being lost to the atmosphere each year.

2.4.1 Factors affecting denitrification

The main environmental factors which affect the process of denitrification are:

(*a*) *Aeration and soil water content*   Denitrification only proceeds when O_2 supply is insufficient to satisfy the microbial demand. In well-drained soils, N volatilization is related to the moisture content. No losses occur when the moisture content is below 60 per cent of the water holding capacity. Above this figure the rate and magnitude of denitrification are correlated directly with the water regime (because of its relationship with soil aeration, see Fig. 2-5).

(*b*) *Nitrate content of soil*   If this is high and poor aeration conditions exist in the soil, the rate of denitrification is high. Thus there can be an enormous loss of combined N from heavily fertilized fields after heavy rain.

(*c*) *Carbohydrate supply*   A high soil carbohydrate level will promote denitrification because it supplies energy for microbial respiration.

(*d*) *Temperature*   Denitrification is markedly affected by temperature. It proceeds very slowly at 2°C with an optimum above 25°C and is still rapid

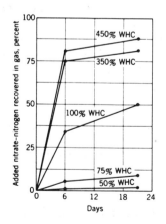
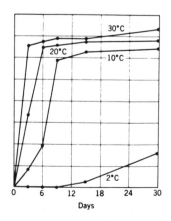

Fig. 2-5   Effect of moisture, expressed as water holding capacity (WHC), and temperature on denitrification in soil to which a known amount of nitrate has been added. The percentage of this nitrate nitrogen given off as gas from the soil indicates the degree of nitrification. (Barea, J.M. and Brown, M.E. (1974). *J. Applied Bacteriol.*, **37**, 583–93.)

at high temperatures up to 60–65°C. Thus, the organisms responsible are mainly thermophilic.

Denitrification, of course, represents a permanent loss of combined nitrogen from the biosphere. There may also be temporary losses which can nevertheless last for a considerable time; for example, when nitrogenous material is deposited from large bodies of water to form sediments below the euphotic regions, thus becoming largely unavailable for reuse by biological systems. Much sewage nitrogen discharged into lakes or the sea, is thought to be temporarily 'lost' in this way.

2.5   Gains of combined nitrogen to the nitrogen cycle

The considerable losses that occur from the global pool of combined nitrogen by processes such as denitrification must obviously be balanced by gains if the nitrogen cycle is to continue functioning. This is, in fact, the case and there are several avenues whereby the depleted supplies of combined nitrogen can be replenished.

2.5.1   Gains from the earth's mantle

Weathering of rocks can release small quantities of combined nitrogen for use by living organisms. The source of this ammonia depends on the type of rock being weathered: in the case of sedimentary rocks the combined nitrogen will have arisen from the remains of organic matter deposited with the particles of the sediments from which the rock was formed; in the case of igneous rocks the nitrogen (as ammonia) will have been trapped within the rock structure from the atmosphere of the earth at the time of solidification. The former source cannot really be considered as a 'new' combined nitrogen source because in this case combined nitrogen removed 'temporarily' from the nitrogen cycle many millions of years ago is being brought back into circulation.

2.5.2   Gains from atmospheric deposition

Nitrogen compounds scrubbed from the atmosphere by rain (wet deposition) or deposited as wind-blown dry organic material (dry deposition e.g. pollen grains), can provide significant additions of combined nitrogen to the soil. Input of combined nitrogen from this source varies considerably with locality, and measurements of from 2 kg to 19 kg N ha^{-1}y^{-1} have been recorded, depending on the presence or absence of heavy industry in the vicinity.

At one time it was thought that lightning flashes were the source of this combined nitrogen, the intense heat and electrical energy of these flashes causing dinitrogen to combine with oxygen or hydrogen in the atmosphere to produce oxides of nitrogen, and ammonia. Today we know that inorganic nitrogen generated by lightning flashes and electrical sparks, forms only a very small part of the combined nitrogen of the atmosphere;

the bulk of it comes from volatilization of nitrogen as a result of grass and forest fires and, in recent years, industrial combustion of fossil fuels, with volcanic activity also providing an input. Thus only about 2 per cent of the combined nitrogen acquired by the soil as a result of atmospheric deposition can be considered as true accession, the remainder being simply recycled combined nitrogen derived mainly from the burning of organic matter.

2.5.3   Gains from biological nitrogen fixation

By far the greatest accession of newly combined nitrogen to the biosphere comes from the fixation of atmospheric N_2 by certain prokaryotic organisms; there is little doubt that the largest proportion of the world's organic nitrogen has arisen from this source. Why the genes for this tremendously important nitrogen fixing ability possessed by these primitive organisms have not been transferred to higher life forms during the course of evolution, is still one of the enigmas of biology. It is estimated that over 100 000 000 tonnes of combined nitrogen are made available to the biosphere each year by this process. In the USA (the world's greatest user of fertilizer nitrogen), this figure represents more than three times the amount of nitrogen added as fertilizer in agriculture.

The organisms responsible for biological nitrogen fixation may be classified according to their mode of nutrition, the two primary groups being the symbiotic and non-symbiotic classes.

Non-symbiotic nitrogen fixation   The organisms responsible for this type of fixation are found free-living in the soil. The energy required for the fixation of nitrogen in these soil organisms comes from either photosynthesis or the oxidation of organic materials absorbed from the soil.

The *photosynthetic organisms* (phototrophs) are of two types. By far the most important are certain genera of the Cyanophyta (blue-green algae) such as *Nostoc* and *Anabaena* which fix their nitrogen in special cells called heterocysts. These algae are particularly important in the wetland cultivation of rice (the staple diet of a large proportion of mankind), where as much as 50 kg ha^{-1} annual crop^{-1} can be fixed and eventually made available for crop growth, thus rendering nitrogen fertilization largely unnecessary.

Certain green photosynthetic bacteria and sulphur photosynthetic bacteria (which derive their electrons from reduced sulphur compounds), and a number of non-sulphur photosynthetic bacteria (e.g. *Chlorobium* spp., *Rhodospirillum* spp.) can also fix nitrogen using light energy.

The *non-photosynthetic* soil organisms (heterotrophs) which fix nitrogen, rely on the oxidation of organic compounds present in the soil to provide the energy for the fixation process. These are bacteria from a number of genera, the better known of which are *Clostridium* (anaerobic), *Azotobacter, Azotococcus*, and *Beijerinckia* (all aerobic).

Symbiotic nitrogen fixation   By far the most effective nitrogen-fixing organisms are those which form symbiotic relationships with higher plants and are able to draw on the carbohydrates produced by these plants to fuel their nitrogen fixing activity and provide carbon skeletons for the production of nitrogenous compounds.

The most important of these relationships, certainly from the agricultural point of view, is the symbiosis that exists between a large number of members of the legume family, Fabaceae (Leguminosae), and several species of the bacterial genus *Rhizobium*. A good stand of clover or lucerne can, for example, fix between 100 and 400 kg N ha^{-1}y^{-1}.

The first evidence that leguminous plants could fix atmospheric dinitrogen was obtained by Boussingault in 1838 when he showed that such plants could increase, rather than deplete, the nitrogen content of the soil in which they were growing. It was only in 1886, however, that Hellriegel and Wilforth hypothesized that the nitrogen fixation was brought about by bacteria living in nodules on the roots of these plants (since confirmed by^{15}N experimentation).

We now know that these bacteria occur free-living in the soil (where they do not fix nitrogen) and are attracted to the roots of young legumes by secretion from these roots. A certain degree of specificity is considered to exist between the legume host species and the bacterial symbiont. For example, *Rhizobium leguminosarum* infects the roots of *Pisum* spp., *Lathyrus* spp. and *Vicia* spp, *R. japonicum*, the roots of the soya bean (*Glycine max*) and *R. phaseoli* the roots of *Phaseolus* spp. Host recognition is apparently assisted by lectins. These are proteins produced by the bacteria which recognize certain sugar residues in the root hair wall of the 'correct' host and allow the bacteria to bind to it (Fig. 2-6). The rhizobia

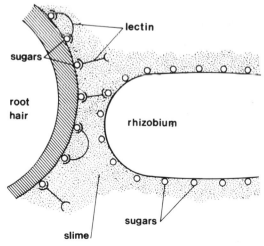

Fig. 2-6   Lectin binding of rhizobium to root hair. (After Kijne, J.W. *et al.*, *Plant Sci. Lett.*, **18**, 65–74.)

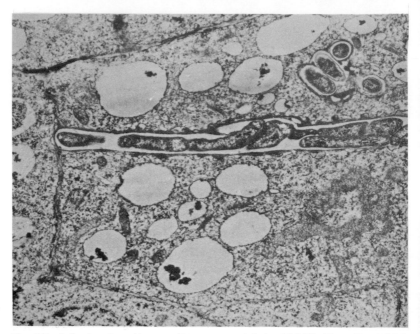

Fig. 2-7 Infection thread in a soya bean root nodule (× 10 000). (From Goodchild, D.J. and Bergensen, F.J. (1966). *J. Bact.*, **92**, 204–13.)

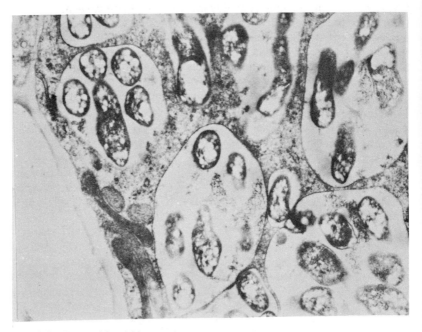

Fig. 2-8 Bacteroids within membrane envelopes in the cell of a mature soya bean nodule (× 23 000). (From Goodchild, D.J., and Bergensen, F.J. (1966). *J. Bact.*, **92**, 204–13.)

synthesize compounds which deform the root hairs of the host, causing them to curl and allowing them to be penetrated by the bacteria. Following infection the root hair wall invaginates and extends into a tube-like structure, thus initiating the 'infection thread' which penetrates the walls of cortical cells and allows the bacteria to migrate into the inner cortex of the root (Fig. 2-7). They are finally released into the cytoplasm of these cells where they multiply and enlarge into swollen, irregularly shaped *bacteroids*. Groups (4 to 6) of these bacteroids become surrounded by membranes produced by the cell to form a number of isolated nitrogen fixing colonies within the cell (Fig. 2-8).

Just prior to the liberation of bacteria from the infection threads, the cortical cells of the host, in the region of infection, multiply rapidly to form the root nodule. Its final structure consists of a central region containing nitrogen-fixing bacteroids, surrounded by a rhizobia-free area into which the vascular tissue of the root has developed (Figs 2-9 and 2-10). The cells of the central core of the nodule possess twice the chromosome number of the normal cells of the host plant; the significance of this phenomenon is still a matter for debate.

On cutting open a functional root nodule, the central region will be seen to have a light pink colour due to the presence of *leghaemoglobin* in the membranes surrounding each bacteroid group. This haem protein is thought to regulate the supply of oxygen to the bacteroids whose nitrogen fixing enzyme, nitrogenase, is readily inactivated by free oxygen.

The bacteroids are able to fix nitrogen, using energy supplied as carbohydrate produced by the host and delivered via the nodule's phloem link with the root vascular system. The bacteroids in turn supply their host (via the xylem link) with nitrogenous products which are produced in excess of their own needs. Thus the symbiosis is a true one, both organisms benefiting from the relationship. It is also a highly specialized one with the host demonstrating a number of features which accomodate and succour its

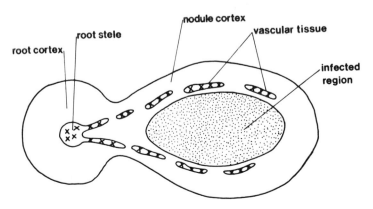

Fig. 2-9   Plan of a typical root nodule.

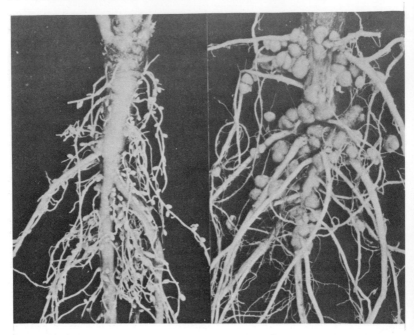

Fig. 2-10   Root nodules on red clover (left) and soya bean (right). (Burton, J.C. and Alexander, M. (1977). *Soil Microbiology*, 2nd ed. John Wiley & Sons, New York.)

symbiont (e.g. nodule structure, diverted vascular strands, leghaemo-globin, and rhizobium-attracting root exudates).

Although ammonia is the first product of nitrogen fixation in the bac-teroids it is rapidly converted to glutamine by the enzyme glutamine syn-thetase (see Chapter 3). Fixed nitrogen may be exported in this form by the nodule to the host plant, but more usually it is converted into compounds such as asparagine, citrulline or the ureides for loading onto the xylem con-nection to the host (see Chapter 6).

In addition to the *Rhizobium*-legume symbiosis a number of other nitrogen fixing symbioses are important in the accession of combined nitrogen to the global nitrogen cycle. Major examples are:

(a) *Actinomycetes with* Alnus *spp. (alder)*, Casuarina *spp. and* Myrica gale (*bog myrtle*)   The actinomycetes, a transitional group of micro-organisms between simple bacteria and fungi, are capable of fixing between 12 and 200 kg N ha^{-1}y^{-1} in the symbioses they form with *Alnus, Casuarina* and *Myrica*.

(b) *Blue-green algae with gymnosperms, ferns, lichens and liverworts* *Macrozamia* and *Encephalartos* (both cycads) have certain of their roots infected with blue-green algae of the genera *Nostoc* or *Anabaena*. These roots adopt negative geotropism and emerge above the ground where the algae, present in a ring of cortical cells near the epidermis, are able to fix nitrogen using photosynthetically derived energy. *Macrozamia* has been shown to be an important contributor of newly-fixed nitrogen in Australian *Eucalyptus-Macrozamia* woodlands. *Podocarpus* and *Cycas* are gymnosperms which produce nodule-like structures on their roots and it is now known that *Podocarpus* nodules contain micro-organisms (as yet uncharacterized) which can be nitrogen fixing. The free-floating fern, *Azolla*, also forms a symbotic relationship with blue-green algae and can fix up to 162 kg N ha^{-1}y^{-1}. In the paddy fields of Indonesia, decomposing *Azolla* contributes a major part of the nitrogen required for the growth of rice crops. Certain lichens and liverworts are also involved in nitrogen fixing symbioses with blue-green algae, but the ecological significance of these associations is considered to be small.

(c) *Nitrogen-fixing bacteria with plant leaves* A number of plants (e.g. *Pavetta* spp., *Psychotria* spp., *Ardisia* spp. and *Gunnera* spp.) produce leaf nodules which become infected with nitrogen-fixing bacteria or blue-green algae. Only in the case of *Gunnera*, however, does the host plant appear to derive any benefit from the association.

(d) *Rhizosphere associations (Rhizocoenoses)* Increasing attention is being paid to the loose association that exists between heterotrophic, non-symbiotic, nitrogen-fixing organisms and the roots of a number of species of higher plants. The contribution of such free-living bacteria to global nitrogen fixation is considered to be relatively small because of the constraint placed on their assimilatory activity through lack of adequate energy resources. Bacteria living in the rhizosphere are, however, able to draw on the considerable energy supply provided by root exudates for their nitrogen-fixing activity. Certain nitrogen-fixing bacteria have been shown to be present in large concentrations on or near the root surface of plants such as maize, wheat, sorghum, rice, sugar cane and a fairly wide range of grasses. There seems to be little doubt that the higher plants concerned benefit greatly from this loose association with rhizosphere nitrogen-fixing bacteria as they grow surprisingly well in the absence of fertilizer nitrogen. The bacteria are special strains of *Azotobacter, Spirillum, Azospirillum* and *Beijerinckia* which appear to be adapted to life in the rhizosphere. Experiments using acetylene reduction techniques indicate that such bacteria could be fixing as much as 100 kg Nha^{-1}y^{-1}, thus the association may be an extremely important agro-economic one. A particularly close relationship exists between *Azospirillum* and its 'hosts' (mainly tropical forage grasses, e.g. *Panicum maximum, Digitaria* spp., and tropically grown maize, wheat, rye and sorghum). Here the bacterium actually infects the root itself and may be found both intracellularly and in intercellular spaces of this organ.

The mechanism of nitrogen fixation This mechanism is a complex one and is, as yet, not fully understood. The overall reaction is as follows.

$$15ATP + 6H^+ + 6e^- + N_2 \rightarrow 2NH_3 + 15ADP + 15Pi$$

(where Pi = inorganic phosphate)

The reaction is catalysed in microorganisms by the enzyme complex, nitrogenase. The process is highly energy-demanding.

The enzyme *nitrogenase* is composed of two iron-sulphur proteins. One is called the Mo-Fe protein and contains two molybdenum atoms, up to thirty-six iron atoms and about the same number of sulphide ions per molecule of approximately 220 000 molecular weight. There are four subunits to each molecule. The other iron-sulphur protein is smaller, with a molecular weight of about 60 000 and consists of two subunits. It is known as the Fe protein and contains one 4Fe–4S cluster per molecule. Both these proteins, especially the latter, are extremely sensitive to the presence of oxygen and will fix nitrogen only in an anaerobic environment.

The nitrogen-fixing reaction requires the substrate, N_2, an electron donor, ATP, ferredoxin (or flavodoxin), nitrogenase and an anaerobic environment in order to operate. The electron source apparently differs between organisms; probable sources are NADH, NADPH and pyruvate. (For *Rhizobium* the donor has not yet been established with certainty.) It is presently thought that the reaction occurs as follows.

An electron is transferred from one of the above electron donors to an electron carrier which can either be ferredoxin or flavodoxin (a flavin complexed protein), depending on the organism. The Fe protein receives electrons from the carrier and then complexes with magnesium-associated ATP (Mg-ATP). The electron transfers to the Mo-Fe protein which has already bound on a N_2 molecule. The transfer is accompanied by the hydrolysis of the Mg-ATP association into Mg-ADP and Pi, providing the energy for a 'super reduced' Mo-Fe protein. In this super-reduced condition the Mo-Fe protein is able to carry out the final reduction of N_2 to NH_3. The reduction process is summarized in Fig. 2-11.

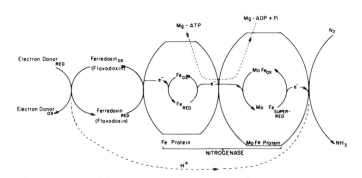

Fig. 2-11   Probable mode of action of nitrogenase.

The process of nitrogen fixation is energetically expensive, requiring the expenditure of 355 kJ per mole of ammonia produced, and in the legume-*Rhizobium* symbiosis can prove a heavy drain on the photosynthetic product of the host plant. Associated with nitrogenase activity in nitrogen-fixing bacteria, is a hydrogenase reaction which reduces protons to hydrogen. This hydrogen-evolving process appears to be unavoidable in a large number of nitrogen fixing bacteria and, because it also requires hydrolysis of the Mg-ATP complex to proceed, is an energy wasting reaction as far as the bacterium is concerned. This hydrogenase activity of the nitrogenase enzyme can be responsible for the loss of 40–60 per cent of the flow of energy as ATP through the nitrogenase system and is therefore an important factor in the determination of the productivity of nitrogen-fixing crop plants.

The measurement of nitrogenase activity   Nitrogenase catalyses the addition of hydrogen to a number of substrates containing triple bonds. A particularly useful reaction of this type is the reduction of acetylene to ethylene ($C_2H_2 + H_2 \rightarrow C_2H_4$), as it can be used to give an approximation of the rate of nitrogenase activity in nitrogen-fixing organisms. In the acetylene reduction assay the material to be tested (soil samples, root nodules etc.) is enclosed in a sealed chamber containing acetylene gas. Samples of the gas are removed from time to time and analysed to ascertain the rate at which acetylene is being transformed into ethylene. These two gases can be easily separated and accurately estimated by the gas-chromatographic method.

Unfortunately the apparent Km (Michaelis constant) values of nitrogenase for its various substrates and the forms they take, differ, and correction factors have to be applied in converting acetylene reduction rates to nitrogen reduction rates. These factors are not always accurately known, consequently the method has to be used with caution. A more accurate method for the determination of nitrogenase activity is to measure the rate of fixation of $^{15}N_2$ gas, but this is a tedious method in comparison with the acetylene reduction assay (see Fig. 3-9).

2.5.4   Gains from artificial nitrogen fixation

Agriculture in developed nations is becoming increasingly dependent on the application of artificially manufactured nitrogen fertilizers to maintain high crop yields. This fertilizer is produced by the Haber-Bosch method whereby hydrogen produced from methane is catalytically combined with dinitrogen under high pressure and temperature conditions. It is a highly energy consuming process and thus an increasingly expensive one, but can account for as much as one third of the annual newly fixed nitrogen accession in countries such as the USA.

In this chapter we have discussed the cycling of combined nitrogen in the biosphere and the losses and gains to this cycle which keep it in a state of equilibrium. On a global scale the movement of nitrogen from one compartment of the cycle to another occurs in massive quantities as can be seen in Table 2-2 which is one of a number of approximations put forward by various ecophysiologists. It can be observed that the amount of 'permanent' loss of combined nitrogen from the biosphere each year (via denitrification) is approximately 160 000 000 tonnes which appears to be slightly exceeded by gains from the atmosphere (biological, atmospheric and industrial fixation) in the region of 176 000 000 tonnes. Without the contribution of industrial fixation, however, this surplus would very likely become a deficit.

Table 2-2   Estimate of global nitrogen transfer. (Delwiche, C.C. and Likens, G.E. (1981). In Some Perspectives of the Major Geochemical Cycles. *Scope 17*, **43**. Wiley & Sons, London.)

	Million tonnes per annum
Biological fixation: land	99
ocean	30
Atmospheric fixation (lightning)	7
Industrial fixation	40
Combustion	18
Fires	50
Denitrification: land	120
oceans	40
Ammonia volatilization	75
Dry deposition NH_3 Wet deposition NH_3	79
Dry deposition NO_3^- and NO_2^- Wet deposition NO_3^- and NO_2^-	34
River runoff: NO_3 Organic nitrogen	35

These transfer figures can be compared with the sizes of the major N pools mentioned at the beginning of the first chapter.

3 The Processing of Inorganic Nitrogen by the Plant

As we have seen in the previous chapter the main forms in which nitrogen becomes available for absorption from the soil by plant roots is as ammonium or nitrate ions. In well-aerated, non-acidic soils the activity of the nitrifying soil bacteria ensures that most of the available nitrogen is present as nitrate, and it is probably true to say that for higher plants in general and crop plants in particular, nitrate is the main source of nitrogen. But there are large areas of the world where soil condition are not conducive to the activity of nitrifying bacteria and here plants rely mainly on ammonium for their nitrogen nutrition. This occurs, for example, in the high-veld grasslands of Zimbabwe with their low pH soils containing few nitrifying bacteria and little nitrate, and the acidic soils of the Western Cape regions of South Africa where plants appear to show a preference for ammonium rather than nitrate as their nitrogen source. Plants can also use other nitrogen sources such as urea and amino acids, although it is doubtful whether these are of much significance under natural conditions, except in the nutrition of saprophytic fungi.

3.1 Absorption of soil nitrate

Most plants are able to extract nitrate rapidly from their root environment, concentrating it in root tissue or in xylem sap at levels far higher than those in the outer medium. For instance, barley plants growing on a culture solution containing 8 mM nitrate have shown nitrate concentrations of over 30 mM in the xylem stream supplying their shoots; similar nitrate concentrating abilities have been shown in sunflower, wheat, maize and *Datura*. Experiments with metabolic inhibitors indicate that the uptake of nitrate is undoubtedly an active process requiring energy to drive it, and the presence of special carrier proteins, nitrate permeases, which catalyse the passage of nitrate ions across cell membranes (and especially those of the root hairs) has been suggested. We still know very little about the details of this process and much work remains to be done in this field.

There is a question associated with the rapid accumulation of nitrate by plants as to how the plant maintains internal electroneutrality in spite of the large influx of negatively charged nitrate ions. The Israeli workers Benzioni, Vaarda and Lips, having observed the inverse ratio that exists in many plants between nitrate and malate accumulation, proposed a solution to this problem. This suggests that malate is produced in the leaves from photosynthate and exported to the root via the phloem in exchange for nitrate arriving in the xylem stream, thus achieving electroneutrality in the leaf. In the root, malate is decarboxylated and the negatively charged bicarbonate ion so produced is exchanged for soil nitrate, avoiding the necessity of importing large amounts of positively charged ions to maintain electroneutrality. The process is illustrated in Fig. 3-1.

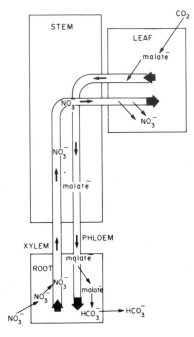

Fig. 3-1 The Benzioni, Vaarda, Lips bicarbonate/nitrate exchange model for nitrate absorption.

Nitrate uptake by plant roots is an inducible process. Plants which have been starved of nitrate absorb this ion only slowly when first reintroduced to it and it takes a few hours for the maximum rate of absorption to be reestablished. This effect is probably due to the time needed for the synthesis of the carrier protein that facilitates nitrate absorption, as experiments with wheat have shown that inhibitors of RNA synthesis prevent the reestablishment of the ability of the root to absorb nitrate after a period of nitrate starvation.

Factors which have an important influence on the absorption of nitrate by plants are:

(*a*) the availability of energy-rich compounds to drive the active permease mechanism;

(*b*) soil temperature (nitrate absorption falls off markedly at low temperatures); and

(*c*) pH (the maximum absorption of nitrate occurs from acidic root growth media). The presence of ammonium ions in the root medium greatly inhibits the uptake of nitrate (see Fig. 3-2 and Table 3-2), but the reason for this is still uncertain.

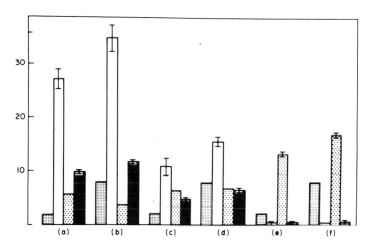

Fig. 3-2 Relationships between nitrogen treatments, nitrate and amino nitrogen concentration in xylem sap, and nitrate reductase (NR) activity in leaves. □, nutrient NO_3^- (mM). ▨, xylem sap NO_3^- (mM). ▨, xylem sap amino-N (mM). ■, NR activity (μmol NO_2 g^1 wt h^{-1}). Nitrogen sources: **(a)** 2mM NO_2^-; **(b)** 8mM NO_3^-; **(c)** 1mM NO_3^- + 1mM NH_4^+; **(d)** 4mM NO_3^- + 4mM NH_4^+; **(e)** 2mM NH_4^+; **(f)** 8mM NH_4^+. (From Lewis, O.A.M., James, D.M. and Hewitt, E.J. (1982). *Ann. Bot.*, **49**, 39–49.)

3.2 Absorption of soil ammonium

The kinetics of ammonium absorption by plants indicate that this process has both active and passive components. Experimentation with metabolic inhibitors in plants such as tomato have shown that the ammonium absorption rate can be halved when respiratory energy release is inhibited, but that uptake is never completely suppressed as is the case in nitrate absorption.

Environmental factors which affect the absorption of ammonium by plants are:

(*a*) pH (many plants absorb ammonium maximally from a medium with a pH of approx. 8);

(*b*) temperature (ammonium absorption by roots falls off at low temperatures but the process is less sensitive to cold than is the absorption of nitrate).

Ammonium uptake rate is also greatly dependent on the availability of a good carbohydrate supply to the root, much more so than is the case in nitrate absorption. This is probably because carbon skeletons are immediately necessary for the producton of organic amino molecules from ammonium which would otherwise build up to toxic levels in the root. Nitrate, being less toxic than ammonium, may be temporarily stored or translocated and need not be reduced until it is assimilated, thus the ready availability of carbon skeletons is less critical when nitrate is being absorbed. Usually, there is no spectacular concentration of ammonium in the xylem stream as in the case of nitrate nutrition, because most plants appear to assimilate absorbed ammonium in their roots and load preformed amino compounds into the xylem, rather than the toxic ammonium ion (see Table 3-2).

In most plants studied, the uptake of ammonium ions is far more rapid than that of nitrate ions.

Table 3-1 Nitrate, ammonium and amino-compound composition (μmol N ml^{-1}) of xylem sap of 20 day old barley seedlings grown on nitrate, nitrate: ammonium 1:1 or ammonium at a nitrogen concentration of 2mM. (From Lewis, O.A.M., James, D.M. and Hewitt, E.J. (1982). *Ann. Bot.*, **49**, 43.)

	NO_3^-	$1:1\ NO_3^-:NH_4^+$	NH_4^+
Aspartate	0.097	0.093	0.143
Threonine	0.113	0.097	0.249
Serine	0.425	0.425	0.741
Asparagine	0.450	0.142	0.680
Glutamate	0.238	0.800	1.844
Glutamine	2.041	2.910	7.122
Proline	–	–	–
Glycine	0.033	0.040	0.053
Alanine	0.133	0.110	0.213
Valine	0.105	0.110	0.251
Cystine	–	0.020	0.015
Methionine	0.006	0.006	0.021
Isoleucine	0.024	0.039	0.058
Leucine	0.026	0.037	0.061
Tyrosine	0.007	0.010	0.006
Phenylalanine	0.016	0.050	0.019
Lysine	0.117	0.237	0.460
Histidine	0.117	0.039	0.168
Arginine	0.328	0.222	0.436
Total organic N	4.28	5.39	12.54
Ammonium N	1.26	0.81	0.65
Nitrate N	27.25	11.01	0.17
Total N	32.79	17.21	13.36

Important features referred to in the text are underlined

3.3　Nitrate vs. ammonium nutrition

The following four sections summarize the advantages and disadvantages of nitrate and ammonium as nitrogen sources for plant growth.

3.3.1　Advantages of nitrate as a nitrogen source

(*a*) The nitrate ion appears to be non-toxic to plants, and certain crop plants and vegetables (notably members of the Chenopodiaceae) can accumulate large concentrations in their tissues. (However, animals can be poisoned by such plants, particularly when storage leads to the accumulation of high nitrite concentrations.)

(*b*) The absorbtion of cations, especially potassium (K), calcium (Ca) and magnesium (Mg), is enhanced with nitrate nutrition. This effect is considered by many workers to be due to the rise in soil pH following the excretion of bicarbonate ions by the plant in exchange for nitrate, thus producing favourable conditions for cation uptake.

(*c*) Although most plants possess nitrate reductase in both root and leaves, crop plants such as maize, barley and wheat assimilate most of the absorbed nitrate in their leaves, especially under conditions of high nitrate availability. This arrangement brings the nitrogen assimilatory processes requiring energy and carbon skeletons into close proximity with the photosynthetic machinery producing these commodities, obviating the complex translocatory pathways necessary in ammonium nutrition.

(*d*) Nitrate has been shown to enhance plant productivity under saline conditions. The reasons for this are not completely clear.

3.3.2　Disadvantages of nitrate as a nitrogen source

(*a*) Before nitrate can be used by the plant, it has to be reduced to NH_4^+, an energy consuming process requiring $347kJ$ mole^{-1} to perform. This represents a significant loss of energy from the plant's overall economy, energy which could otherwise have been used in increasing the plant's productivity.

(*b*) Perhaps the most serious disadvantage of the use of nitrate as a fertilizer is the ease with which it may be leached from soils. This is due to the negative charge on the ion, which prevents it being retained by most soils whose particles also possess negative charges. This factor, together with the ease with which denitrification of NO_3^- may take place under anoxic conditions, is responsible for the major loss of nitrogen from fertilized fields.

(*c*) Nitrate absorption is an active, energy dependent process requiring ATP to drive the permeases responsible for the uptake of the ion from the soil. Anoxic conditions such as those induced by waterlogging of the soil can thus severely inhibit root absorption of nitrate through their inhibition of oxidative phosphorylation.

(*d*) Iron and certain trace element deficiencies can be induced by nitrate nutrition. This effect is probably due to the internal binding of the metals with organic acids which are produced in large quantities in root and stem as a result of nitrate feeding. (This effect can be particularly serious in maize.) The flux of negatively charged nitrate into the plant stimulates the production of positively charged hydrogen ions in the form of organic acids, apparently to maintain electrical neutrality.

3.3.3   Advantages of ammonium as a fertilizer

(a) Unlike nitrate, ammonium does not require reduction prior to utilization by the plant, thus resulting in considerable energy saving. A major agricultural advantage enjoyed by ammonium feeding of plants over nitrate feeding is the more effective retention of this ion by the soil, because of attractive forces that exist between the positively charged soil ammonium ions and the negatively charged soil particles (see Chapter 2). Nitrate ions, being negatively charged, are easily leached from the soil by rain and consequently there is an enormous loss of combined nitrogen in this form from agricultural lands – in fact, it is widely considered that over 50 per cent of the increasingly expensive nitrogen fertilizer applied by farmers to the land is simply washed out to sea, often causing serious river and dam pollution on the way. It is mainly for this reason that serious efforts are now being introduced to prevent or retard the conversion of ammonium fertilizer to nitrate in the soil by the application of chemicals such as nitrapyrin which selectively inhibit the activity of the nitrifying bacteria, *Nitrosomonas* (see Chapter 9).

(b) Ammonium has apparently a double absorption system, one energy dependent and one energy independent. Thus, metabolic inhibitors and anoxia of the root environment have relatively little effect on ammonium absorption compared with nitrate absorption.

(c) Ammonium nutrition enhances anion absorption, particularly phosphate. Enhanced phosphate absorption is due to a lowering of rhizosphere pH by H^+ excretion in response to ammonium absorption, resulting in a conversion of $H_2PO_4^-$ to HPO_4^{2-} ions, which are absorbed several times faster than $H_2PO_4^-$.

3.3.4   Disadvantages of ammonium as a nitrogen source

(a) The ammonium ion can be toxic to plants. This toxicity appears to have two main causes:

(i) ammonium uncouples photophosphorylation at concentrations as low as 2 mM, thus severely restricting ATP production in leaves. It is probably for this reason that nutrient ammonium assimilation takes place in the root and that ammonium is only loaded onto the xylem supply to the leaf in small quantities;

(ii) ammonium absorption is electrically balanced by the excretion of H^+ by the root into the soil. Acidification of the root environment severely retards growth, resulting in stunted rooting systems and greatly impaired nutrient absorption. This effect can be largely overcome by the liming of soil or the addition of calcium carbonate ($CaCO_3$) to nutrient solutions.

In most plant tissues the enzyme glutamine synthetase is present which produces the amide glutamine from glutamic acid and ammonium. This enzyme acts as an ammonium 'detoxifier' in tissues containing excessive quantities of the ion; this was considered to be its main function in the plant until recently (see Section 3.6).

(b) Because of the immediate need to combine ammonium organically after absorption to prevent toxicity, large quantities of carbohydrate are immobilized in the production of N compounds such as glutamine and asparagine. At high levels of ammonium nutrition, this can severely restrict the amount of material available for structural purposes, resulting in small, weak plants and slow growth.

(c) Ammonium nutrition suppresses the absorption of K, Ca, Mg and NO_3^-, probably because they share a common absorptive permease system.

(d) In most plants, absorbed nitrate is processed in both root and shoot, but nutrient ammonium appears to be assimilated exclusively in the root. As mentioned previously, relatively little is loaded into the xylem for translocation to the shoot; instead, large quantities of amides and amino acids are transported to the shoot to maintain its nitrogen supply (see Fig. 3-2). This exclusive assimilation of nutrient ammonium in the plant root must inevitably place stress on the shoot-root translocation system of the plant which is obliged to provide the root with large quantities of carbon skeletons manufactured in the leaf to utilize the ammonium absorbed.

3.4 Response of the plant kingdom to nitrate-ammonium nutrition

The answer of the plant kingdom to the limitations on growth imposed by the different forms of nutrient nitrogen has been to develop three types of plant with respect to their nitrogen response. These forms are (a) nitrate nutrition adapted species, (b) ammonium nutrition adapted species and (c) ammonium + nitrate adapted species.

3.4.1 Nitrate adapted species

These plants usually occur on neutral or alkaline soils where the soil pH is conducive to the activity of the soil nitrifying bacteria which convert all or most of the available soil ammonium into nitrate. Such plants are termed *calcicoles* and many are seriously damaged, or killed, by high levels of soil ammonium. Many species of *Euphorbia* from the drier regions of South Africa and plants like *Scabiosa* spp., *Cucumis* spp. and *Datura* spp. are examples of calcicoles ('at home' in lime soils).

3.4.2 Ammonium adapted species

In those parts of the world where soils are acid, the activity of the nitrifying bacteria is inhibited and the main form of available nitrogen is the ammonium ion. Many plants growing under these conditions show a much better response to ammonium nutrition than nitrate nutrition and are termed *calcifuges* ('fleeing from lime'). A large number of plants belonging to the Ericaceae show a preference for the ammonium rather than the nitrate ion. Climax species of grasslands in many tropical and sub-tropical parts of the world are apparently dominated by calcifuge species as many of the grasses show a definite preference for ammonium nutrition and grow in acidic soils containing little nitrate for most of the year. There is evidence that the roots of these plants actually excrete inhibitors of nitrifying bacteria to prevent nitrate formation (and its subsequent loss by leaching).

The preference for ammonium or nitrate nutrition by individual plant species appears to be dependent on the habitat in which each species has evolved. For example, of four species of conifer studied by Krajina and his colleagues, two grew best on nitrate, one preferred ammonium and one preferred ammonium or a combination of ammonium and nitrate. The two species which grew best on nitrate are found growing naturally in non-acidic soils where nitrification takes place readily, particularly if the soils are alkaline and rich in lime. The other two species are found in acidic soils where nitrification is slow or absent.

The physiological differences between calcicole and calcifuge plants have not been fully investigated and further research in this field is still necessary.

3.4.3   Plants adapted to ammonium + nitrate nutrition

The majority of plants, particularly the agriculturally important crop plants, fall into this category. They show a better growth response when fed a mixed ammonium and nitrate nitrogen source than when fed one of these sources alone. Cox and Reisenauer in California discovered the exciting fact that wheat can increase its growth rate and productivity by over 50 per cent when certain combinations of nitrate + ammonium are provided, compared with the feeding of the N sources alone. Higher protein production in plants provided with both nitrogen sources as opposed to one source, has also been shown in maize, soya bean, tobacco and flax, The optimum ammonium:nitrate ratio differs between plant species and plant.

The cause of this enhancement of plant growth and productivity by mixed N sources is still largely unknown, but it can be hypothesized that it is due to a combination of the favourable features of both nutrition sources and a mutual 'cancelling out' of some of the unfavourable factors.

With the recent introducton of nitrification inhibitors, such as nitrapyrin, carbon bisulphide (CS_2) and the water soluble tri-thiocarbonates, it is now possible to partially regulate the conversion of ammonium to nitrate in agricultural fields. Thus, new scope has been provided for more efficient utilization of fertilizer nitrogen by the controlled application of a mixed source N together with a nitrification inhibitor, although more research is required in this area.

3.5   The assimilation of nitrate

Once nitrogen has been absorbed by the plant, it is reduced to ammonium before assimilation into organic compounds can take place. This process is called assimilatory nitrate reduction, as opposed to respiratory nitrate reduction which takes place in certain soil bacteria that use nitrate as an electron acceptor in place of molecular oxygen.

It is now widely accepted that the process of nitrate reduction to ammonium is mediated by the two enzymes, nitrate reductase (converting nitrate to nitrite), and nitrite reductase (converting nitrite to ammonium), which operate sequentially. This reduction requires a great deal of energy as the following reaction shows.

$$NO_3^- + H_2O + 2H^+ \longrightarrow NH_4^+ + 2O_2 - 83 \text{ Kcal (347 kJ)}$$

3.5.1 Nitrate reduction

The main sites of nitrate reduction in the plant are the leaf and the root, usually with a much greater reducing activity evident in the leaf, particularly in crop plants such as barley, wheat and maize.

Some plants, for example, cocklebur (*Xanthium pennsylvanicum*), appear to reduce nitrate solely in their leaves, as no nitrate reductase activity has been found in their roots. The inability to demonstrate nitrate reductase activity in a tissue by assay methods cannot, however, be taken as absolute proof of its absence as there are a number of substances which, once released from the cell following enzyme extraction, inhibit or destroy the activity of this enzyme. For instance, until recently it was considered that the sunflower (*Helianthus annuus*) reduced nearly all of its nitrate in the root, as very little trace of nitrate reductase activity could be detected in the leaves. When suitable protecting agents against proteolytic enzyme activity and polyphenolic inhibition (casein and polyvinylpyrrolidone) were added to the leaf enzyme extract, nitrate reductase activity five times greater than that present in the root was detected.

In most higher plants it seems that nitrate, once absorbed by the plant root, crosses the root cortex where some of it is reduced, the remainder being loaded onto the xylem stream for reduction in the leaf. The partitioning of nitrate reduction between root and shoot could depend on the availability of nitrate in the root environment. In crop plants, nitrate in the form of applied fertilizer is usually present in high concentration in the soil at the seedling stage and is absorbed in such quantity that only a small fraction can be reduced in the root, the bulk being processed by the leaf. Late in the growing season when soil nitrate supplies are depleted by leaching and by plant absorption, it is quite probable that the root becomes the main site of nitrate reduction, as its nitrate reductase activity is now sufficient to cope with the limited nitrate uptake. It must be remembered that nitrate is not the only form of nitrogen exported by the root to the shoot: the root also supplies the shoot with a number of amino compounds which it manufactures for export via the xylem stream (see Chapter 7 and Fig. 7-3).

Nitrate reductase (International Union of Biochemists' classification: E.C. 1.6.6.1), originally isolated from *Neurospora* by Mason and Evans in 1953, is a soluble flavoprotein enzyme and there is much evidence to indicate that it is located on the outer membranes of chloroplasts in the leaf, and colourless plastids in the root. It appears to consist of two subunits which may be repeated; the molecular weight is thus variable between plant species, ranging from 200 000 daltons (fungi and higher plants) to 460 000 daltons (certain green algae). The enzyme contains flavin adenine dinucleotide (FAD), cytochrome b_{557} and molybdenum as part of its electron transport chain. One subunit is involved in the transfer of electrons from reduced nicotinamide adenine dinucleotide (NADH) to FAD and the other in the transfer of electrons from reduced FAD to nitrate via cytochrome b and molybdenum (see Fig. 3-3). In photosynthetic eukaryotic plants NADH is considered to be the preferred reductant whereas in fungi, reduced nicotinamide adenine dinucleotide phosphate (NADPH) appears to be the better source of reducing power.

In prokaryotic photosynthesizing plants (e.g. blue-green algae), the nitrate reductase molecule has a molecular weight of only 75 000 daltons and contains no flavin or cytochrome b. It appears to require reduced ferredoxin as its hydrogen donor.

The overall reaction catalysed by nitrate reductase is as follows.

$$NO_3^- + NAD(P)H_2 \xrightarrow{+2e^-} NO_2^- + NAD(P) + H_2O$$

In addition to catalysing the above reaction, nitrate reductase also possesses 'diaphorase' capabilities, that is, it can transfer electrons from NADH to ferricyanide and various synthetic dyes.

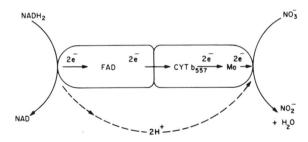

Fig. 3-3 Operation of nitrate reductase.

Nitrate reductase is a substrate-inducible enzyme. In the absence of its substrate the enzyme is present in roots and shoots at very low levels of activity. On the introduction of nitrate to nitrate-starved plants the formation of the enzyme is induced and after four to five hours maximum activity can be achieved in many plants (see Fig. 3-4). The induction of nitrate reductase appears to be caused by the flux of nitrate into the tissue rather than the actual presence of nitrate (which may be stored in vacuoles away from the active site of the enzyme) and is due to the resynthesis of the

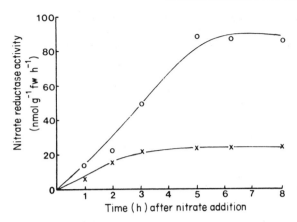

Fig. 3-4 Induction of nitrate reductase in *Datura stramonium* leaves at two nitrate feeding levels: o – o, 10 mM NO_3^-; and x – x, 2 mM NO_3^-. Prior to time 0 plants receive a NO_3^- free nutrient source. (From Probyn, T.A. (1978). MSc Thesis, Univ. of Cape Town.)

enzyme rather than its reactivation. In the absence of nitrate, nitrate reductase activity is rapidly lost, the enzyme having a half-life of some four to five hours (in maize).

Light also acts as an inducer of nitrate reductase in leaf tissue. This effect becomes obvious when leaves receiving nitrate but growing in darkness or shade are exposed to illumination. The level of nitrate reductase activity increases rapidly, but will fall again if the illumination is switched off (see Fig. 3.5). The light effect is no doubt partially responsible for the diurnal rhythm of nitrate reductase (and other enzyme) activity which has been

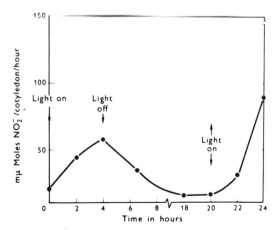

Fig. 3-5 Effect of light on nitrate reductase activity in radish at an illumination of 225 lux. (From Beevers, L. *et al* (1965). *Pl. Physiol.*, **40**, 691-98.)

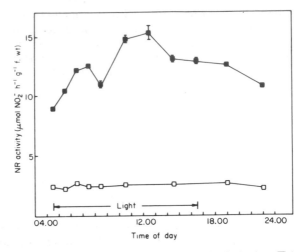

Fig. 3-6   Diurnal rhythm of nitrate reductase activity in barley. ■ leaf; □ root. (From Lewis, O.A.M., Watson, E.F. and Hewitt, E.J., (1982). *Ann. Bot.*, **49**, 31–37.)

observed in a number of plants and is shown for barley in Fig. 3-6. The mechanism for this effect is not properly understood, but could be due to the fact that light increases the plant's capacity for protein and hence nitrate reductase synthesis by providing the necessary carbon skeletons and chemical energy. However, as long as the enzyme is present and a carbohydrate energy supply is available, nitrate reduction can proceed in the absence of light.

There has been extensive discussion over the origin of the NADH used as the electron donor in the reduction of nitrate by nitrate reductase in higher plants. Most evidence points to it being the product of the respiratory oxidation of photosynthate, although some workers consider that it is produced in leaves by photorespiration.

In a large number of plants the activity of nitrate reductase is far lower than that of any of the other enzymes of nitrogen assimilation (see Fig. 3-10), therefore, it can be regarded as the rate-limiting enzyme of those plants which rely on nitrate for their nitrogen source.

This has led to the proposal (which is not universally accepted), that the relative efficiency of a plant as a food protein producer can be determined by assaying its nitrate reductase activity. Recent work by Cresswell and his colleagues in Johannesburg has, however, indicated the possibility that our assay systems for nitrate reductase are faulty and that we may be measuring the activity of only part of the nitrate reducing system.

3.5.2 Nitrite reduction

It is now widely accepted that the nitrite produced in plants by nitrate reduction is immediately further reduced to ammonium by the enzyme nitrite reductase, as nitrite is rarely found free in plants growing under normal conditions. The activity of this enzyme was first demonstrated unequivocably by Hageman, Cresswell and Hewitt at Long Ashton in 1962. It mediates a highly reducing reaction involving the transfer of 6 electrons from ferredoxin to nitrite as follows.

$$NO_2^- + 6 \text{ ferredoxin}_{red} + 8H^+ \xrightarrow{\;6e^-\;} NH_4^+ + 6 \text{ ferredoxin}_{ox} + 2H_2O$$

There has been a long search for intermediates between nitrite and the end product of the reaction, ammonium (e.g. hydroxylamine, hyponitrite), but these have never been found free and it is now considered that the entire reaction occurs on the enzyme without the release of nitrogen compounds of intermediate redox states.

Nitrite reductase (I.U.B. classification: E.C. 1.7.7.1) has been purified in a number of laboratories and its structure has been determined as an iron porphyrin (siroheme) containing protein with a molecular weight of 60 000 to 70 000 daltons.

It has recently been established that the enzyme also contains an iron-sulphur centre (4Fe–4S) as has already been found in sulphite reductase.

Spectrophotometric studies show that both the iron-sulphur centre and siroheme after being reduced with dithionite can be reoxidized by the addition of nitrite, indicating that both these complexes are active in the reduction of nitrite to the ammonium ion.

The reducing reaction in leaves is envisaged as taking place as shown in Fig. 3-7.

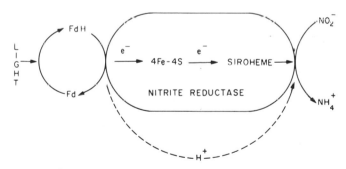

Fig. 3-7 Operation of nitrite reductase.

The electron donor for the nitrite reducing reaction has been the subject of exhaustive investigations. While reduced ferredoxin produced by photosynthesis has been strongly indicated as the electron donor in leaves, the reductant in roots and non-green tissue is still unknown. A possibility is that NADPH, produced by the pentose phosphate pathway, carries out the reduction through some as yet unknown mediator molecule; NADPH cannot function as the direct electron donor.

The validity of the pathway of nitrate reduction to ammonia via nitrate reductase and nitrite reductase is now widely accepted, but recent demonstration of the evolution of nitrogen oxides during nitrite reduction is considered by some to indicate the presence of a second nitrate reducing pathway in plants, the details of which still remain to be investigated.

Partitioning of cell contents by ultracentrifugation and subsequent assay of the separated fractions for nitrate reduction activity indicates that the enzyme is located within the chloroplasts (convenient for reaction with photosynthetically reduced ferredoxin) attached, . possibly, to thylakoid membranes. In roots, nitrite reductase activity is associated with plastids present in the cytosol. Nitrite reductase associated with fungi is a much larger molecule than that found in algae and higher plants, having a molecular weight of 290 000 daltons. It uses NADH or NADPH as electron donors but requires FAD for maximum activity.

Nitrite reductase, like nitrate reductase, is a substrate-inducible enzyme. Nitrite reduction is prevented by anaerobic conditions and the roots of plants like barley actually excrete nitrite if the soil becomes waterlogged and therefore oxygen deficient. Nitrite excretion by the roots of cereal plants also takes place in water culture and nutrient film culture unless the feeding solutions are highly aerated. This indicates that in the root, nitrite assimilation is dependent on reductants produced by aerobic respiration.

3.6   The assimilation of ammonia

This important stage in the cycling of combined nitrogen in the biosphere (discussed in Chapter 2) results in the incorporation of inorganic nitrogen into organic molecules. It takes place mainly in the leaf chloroplasts or root plastids where the enzymes responsible for the process are located.

3.6.1   The glutamate dehydrogenase pathway

The elucidation of the pathway of ammonia assimilation in plants makes an interesting story. Between 1959 and 1962 a major advance in the understanding of this process was the demonstration by Folkes and Sims of the reductive amination of 2-oxoglutaric acid (α-ketoglutaric acid) to glutamic acid in yeast cells (*Candida utilis*) using kinetic ^{15}N labelling techniques. In this reaction ammonia becomes attached to the Kreb's Cycle intermediate, 2-oxoglutaric acid, forming glutamic acid, with NADH acting as the hydrogen donor. The reaction is catalysed as follows by the enzyme glutamate dehydrogenase (GDH, classified by the International Union of Biochemists as EC. 1.4.1.3 L-glutamate: NAD oxidoreductase):

$$
\begin{array}{l}
COOH \\
| \\
CH_2 \\
| \\
CH_2 \\
| \\
C=O \\
| \\
COOH
\end{array}
+ NH_3 + NADH_2
\xrightarrow{\substack{\text{Glutamate}\\ \text{dehydrogenase}\\ \text{(GDH)}}}
\begin{array}{l}
COOH \\
| \\
CH_2 \\
| \\
CH_2 \\
| \\
CH-NH_2 \\
| \\
COOH
\end{array}
+ NAD + H_2O
$$

Because of the almost universal occurrence of glutamate dehydrogenase in plants (localized mainly in the mitochondria but with an NADPH dependent isozyme also present in chloroplasts), it was accepted for many years that reductive amination involving this enzyme was the main portal of entry of ammonia into organic metabolism throughout the plant kingdom.

3.6.2 Glutamine synthetase – glutamate synthase pathway

In the early 1970s, the validity of the prominence of the glutamate dehydrogenase pathway in plant ammonia assimilation came into question. The K_m (Michaelis constant) of glutamate dehydrogenase for ammonia was found to be about 5 mM (although some algae, notably *Caulerpa* spp., have been shown to possess glutamate dehydrogenase with lower K_m's for ammonia), which is high, indicating a relatively low affinity of the enzyme for its substrate. As the concentration of ammonia in higher plant tissue never approaches this level and photosynthesis in plant leaves is uncoupled (i.e. ATP production ceases) at ammonia concentrations in the chloroplast of 2 mM and above, it was difficult to comprehend how the reductive amination reaction could proceed fast enough to satisfy the requirements of amino acid assimilation in the plant. In addition, experiments involving the pulse-feeding of ammonia to *Klebsiella* cultures and the feeding of ^{15}N nitrate to the leaves of higher plants appeared to indicate that newly absorbed or assimilated ammonia was incorporated initially into the amido group of the amide, glutamine, with only a secondary enrichment of glutamic acid taking place. This introduced the possibility of another enzyme, glutamine synthetase, being involved in the assimilation of ammonia. This enzyme, well known in animals, has also been shown to exist in bacteria and the cytosol and plastids of eukaryotic plants. In pea leaf chloroplasts, O'Neal and Joy have demonstrated the enzyme to possess a K_m for ammonia of 0.02 mM, a far more suitable K_m for rapid assimilation of ammonia than that possessed by glutamate dehydrogenase. The enzyme catalyses the following reaction:

$$
\begin{array}{c}
\text{COOH} \\
|\\
\text{CH}_2 \\
|\\
\text{CH}_2 \\
|\\
\text{CHNH}_2 \\
|\\
\text{COOH}
\end{array}
+ \text{NH}_3 + \text{ATP}
\xrightarrow{\text{Mg}^{2+}}
\begin{array}{c}
\text{CONH}_2 \\
|\\
\text{CH}_2 \\
|\\
\text{CH}_2 \\
|\\
\text{CHNH}_2 \\
|\\
\text{COOH}
\end{array}
+ \text{ADP} + \text{Pi}
$$

Glutamic acid   Glutamine

The enzyme requires magnesium as a cofactor, and an energy source in the form of ATP, thus making the reaction virtually irreversible. It should be noted that the newly assimilated ammonia is transformed into the amido group of the glutamine molecule.

While the use by the plant of glutamine synthetase for the initial step in ammonia assimilation now became an attractive possibility, the question still remained as to how the amido nitrogen of glutamine could enter amino acid metabolism, for there was no known enzyme in higher plants which could bring this about. That this reaction did actually occur in plants was shown beyond doubt by Lewis and Pate who fed pea leaves separate solutions of ^{15}N glutamic acid, ^{15}N nitrate and ^{15}N (amide) glutamine and found that the ^{15}N label was incorporated into leaf amino acids in approximately the same proportions regardless of ^{15}N source (see Table 3-2).

Table 3-2   The relative incorporation of ^{15}N into the amino acids of pea leaves, 10h after feeding labelled substrates. (Lewis, O.A.M. and Pate, J.S. (1973). *J. exp. Bot.* **24**, 602.)

| Amino acids | Labelled substrates | | |
	$^{15}NO_3$	^{15}N-glutamic acid	^{15}N (amide)-glutamine
Glutamyl-*	3.76†	4.56	3.43
Serine	2.28	2.70	1.76
Aspartyl-*	2.24	3.25	2.79
Glycine	1.55	2.94	1.94
Alanine	1.44	2.84	1.72
Threonine	1.27	1.80	1.16
Isoleucine	1.13	1.46	0.99
Phenylalanine	0.82	1.73	1.02
Arginine	0.69	1.80	1.31
Valine	0.72	1.19	0.82
Leucine	0.91	1.26	0.81
Tyrosine	0.94	1.32	0.97
Lysine	0.65	1.16	0.95
Histidine	0.65	0.80	0.75

* Includes amino nitrogen of acid + amide but not amido-nitrogen of amide
† Atom percentage excess^{15}N. (Substrates fed at 95 atom percentage excess ^{15}N)

Lea and Miflin at Rothamsted were eventually able to provide the 'missing' enzyme when they demonstrated the presence of glutamate synthase (GOGAT) in chloroplast extracts in 1974. This enzyme, originally discovered in bacteria by Tempest, Meers and Brown, catalyses the reductive transfer of the amido group of glutamine to 2-oxoglutaric acid (α-ketoglutaric acid), producing two molecules of glutamic acid from one of 2-oxoglutaric acid and one of glutamine.

$$
\begin{array}{l}
\text{CO}-\text{NH}_2 \\
| \\
\text{CH}_2 \\
| \\
\text{CH}_2 \\
| \\
\text{CH}-\text{NH}_2 \\
| \\
\text{COOH}
\end{array}
\; + \;
\begin{array}{l}
\text{COOH} \\
| \\
\text{CH}_2 \\
| \\
\text{CH}_2 \\
| \\
\text{C}=\text{O} \\
| \\
\text{COOH}
\end{array}
\; + \; \text{NAD(P)H}_2 \;
\xrightarrow[]{\text{Glutamate synthase}} \;
2
\begin{array}{l}
\text{COOH} \\
| \\
\text{CH}_2 \\
| \\
\text{CH}_2 \\
| \\
\text{CH}-\text{NH}_2 \\
| \\
\text{COOH}
\end{array}
\; + \; \text{NAD(P)}
$$

Glutamine   2-oxoglutaric   Glutamic acid
                     acid

The enzyme remained undetected in plant leaves because the isozyme occurring in chloroplasts is ferredoxin-dependent, and it was only when Lea and Miflin used ferredoxin as a reducing source that the presence of GOGAT in leaves could be demonstrated. In roots, GOGAT has been located in plastids where it is NADH dependent.

The discovery of ferredoxin-dependent GOGAT allowed Lea and Miflin to put forward the GS-GOGAT pathway for the assimilation of ammonia in plants. This pathway is now widely accepted as the major pathway whereby inorganic nitrogen in the form of ammonia is incorporated into plant amino acid metabolism, and is outlined in Fig. 3-8. The product of the reaction, glutamic acid, is an excellent donor of amino nitrogen in transaminating reactions whereby other amino acids are formed from their 2-oxo homologues, thus regenerating the 2-oxoglutaric acid necessary for the GOGAT reaction (see Chapter 4).

Numerous experiments confirming the validity of the GS GOGAT pathway have been performed. Particularly convincing are those involving the use of radioactive isotope of nitrogen, ^{13}N (which are difficult to carry out

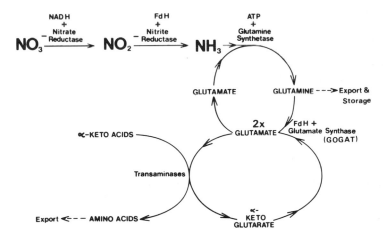

Fig. 3-8 The glutamine synthetase-glutamate synthase (GS-GOGAT) pathway of nitrogen assimilation in plant leaves.

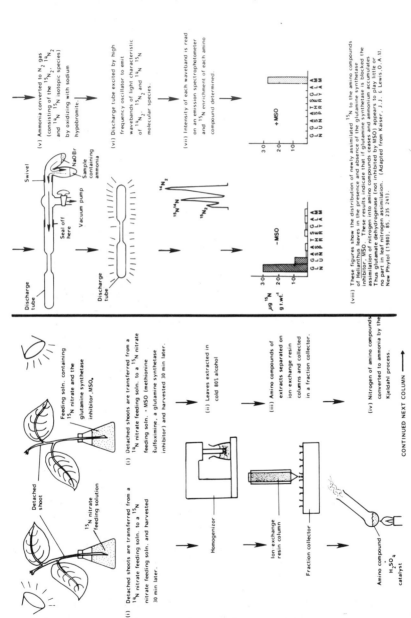

Fig. 3-9 An experiment to illustrate the use of an enzyme inhibitor and the heavy isotope of nitrogen, ^{15}N, in following the pathway of nitrogen assimilation in *Helianthus* leaves.

because of the short half-life of 10 minutes of the isotope) and the heavy isotope of nitrogen, ^{15}N, which requires the use of a mass spectrometer or atomic emission spectrometer to detect its presence. One experiment involving the use of ^{15}N and a selective inhibitor of the enzyme glutamine synthetase is outlined in Fig. 3-9.

The chain of events involved in the assimilation of nitrate nitrogen, and the cellular localities in which each event occurs, is shown in Fig. 3-10.

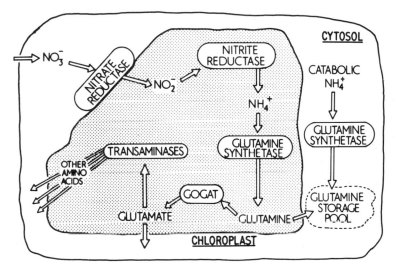

Fig. 3-10   The localization of the enzymes of nitrogen assimilation in a leaf chlorenchyma cell.

The role of glutamate dehydrogenase (GDH) in nitrogen assimilation in the higher plant is now uncertain. Experiments such as that outlined in Fig. 3-9 indicate that it has no involvement at all in this process, yet its activity in plants as measured by *in vitro* techniques can be pronounced, particularly in root tissue (see Fig. 3-11). It is now considered possible that its function in higher plants is the reverse of nitrogen assimilation, that is, it functions to deaminate glutamic acid to produce 2-oxoglutaric acid. This could be a feed-back mechanism to ensure a minimum level of oxoglutarate in the plant for the effective functioning of the Kreb's respiratory cycle, when GOGAT activity is high; it would also explain why GDH activity appears to be much higher in root tissue (with metabolism dependent on respiratory energy) than in leaf tissue (with metabolism dependent on photosynthetic energy).

There is, of course, no doubt that in certain organisms such as *Candida* the enzyme does operate in an assimilatory fashion (i.e. glutamic acid production) but these seem to be special cases where the cell environment is rich in ammonium.

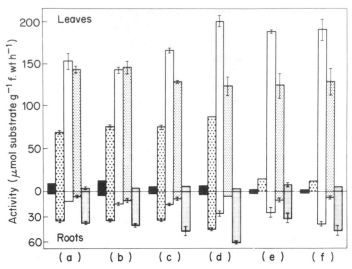

Fig. 3-11 Activities in roots and leaves of nitrate reductase (■), nitrite reductase (⊡), glutamine synthetase (□), GOGAT (⊡) and glutamate dehydrogenase (▦), in relation to nitrogen sources: **(a)** 2mM NO_3^-; **(b)** 8mM NO_3^-; **(c)** 1mM NO_3^- + 1mM NH_4^+; **(d)** 4mM NO_3^- + 4mM NH_4^+; **(e)** 2mM NH_4^+; **(f)** 8mM NH_4^+. (From Lewis, O.A.M., James, D.M., Hewitt, E.J. (1982). *Ann. Bot.*, **49**, 39–49.)

3.6.3   Glutamine synthetase

The enzyme glutamine synthetase (GS), is classified by the International Union of Biochemists as EC. 6.3.1.2 L-Glutamate: ammonia ligase (ADP). The enzyme is a large and complex molecule. In *Escherichia coli* it has a molecular weight of 600 000 daltons and is composed of twelve identical subunits (see Fig. 3-12). In the higher plant there appear to be two isoforms of the enzyme, one occurring in the cytosol (GS_1) and the other in the plastids (GS_2). They have different heat stabilities, pH optima, glutamic acid affinities and regulatory properties. The cytosol isozyme is probably concerned with the assimilation of absorbed ammonia and the reassimilation of ammonia liberated by cell metabolic processes (e.g. photorespiration), whereas the plastid isozyme is directly involved in the assimilation of ammonia produced from the reduction of nitrate.

3.6.4   Glutamate synthase

This enzyme is classified by the IUB as E.C. 2.6.1.53., L-glutamine 2-oxoglutarate aminotransferase (NADPH oxidizing), hence the acronym GOGAT. In bacteria, GOGAT has been shown to have a molecular weight of 800 000 daltons and to be composed of eight subunits, i.e. four each of two types with molecular weights of 135 000 and 53 000 respectively. It is a flavoprotein, containing labile sulphur and iron (not associated with

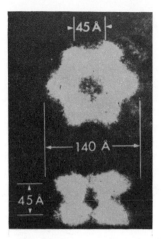

Fig. 3-12 Electron photomicrograph of the enzyme, glutamine synthetase, face and edge views (× 3 160 000). (From Valentine, R.C., Shapiro, B.M. and Stadtman, E.R., *Biochemistry,* 7, 2145.)

haem). The ferredoxin-dependent GOGAT found in chloroplasts is much smaller with a molecular weight of 145 000 daltons. In *Vicia faba* it has a Km for glutamine of 0.3 mM and a Km of 0.15 mM for oxoglutarate. GOGAT does not appear to be present in the cytosol of higher plants, but is restricted to the plastids.

4 Amino Acids

In the previous chapter we have seen how inorganic nitrogen, absorbed by plants from the soil, is transferred into organic nitrogen in the form of glutamine and glutamate. In the following three chapters (Chapters 4 to 6) we will consider some of the more important nitrogenous compounds of the plant which are elaborated by various processes from the original products of the GS-GOGAT pathway, commencing with the amino acids, those essential building blocks from which proteins are made.

4.1 Structure and properties of amino acids

As the name implies the amino acid molecule contains a basic amino group and a carboxylic acid group within its structure, giving amphoteric properties to the molecule (i.e. it can act as a weak base towards a strong acid, or a weak acid towards a strong base). Most of the amino acids occurring in nature are of the α-amino type; the amino group is attached to the α-carbon atom relative to the carboxylic acid group. They differ from one another in the nature of the side chain (R— below).

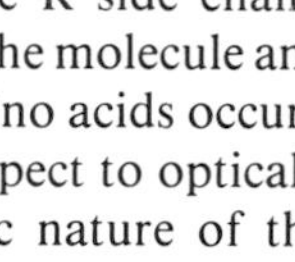

$$R - \underset{\underset{H}{|}}{\overset{\overset{NH_2}{|}}{C}} - COOH$$

The α (i.e. 2—) carbon atom is bonded to four different groups (except in the case of glycine where the R side chain is simply a hydrogen atom) thereby giving asymmetry to the molecule and the ability to rotate the plane of polarized light. All the amino acids occurring naturally in proteins have the L– configuration with respect to optical rotation.

Because of the amphoteric nature of the molecule, amino acids can develop charges of opposite character when they are exposed to acid and alkaline conditions respectively. This is due to the 'Zwitterion Effect' which occurs in the ionized form of the molecule with which the unionized form is in equilibrium in solution. In the ionized form a hydrogen ion switches from the carboxylic acid group (giving it a negative charge) to the amino group (giving it a positive charge), thus producing a dipolar molecule. The effect of acids and bases on this molecule is to produce positively charged and negatively charged molecules respectively in the following reactions.

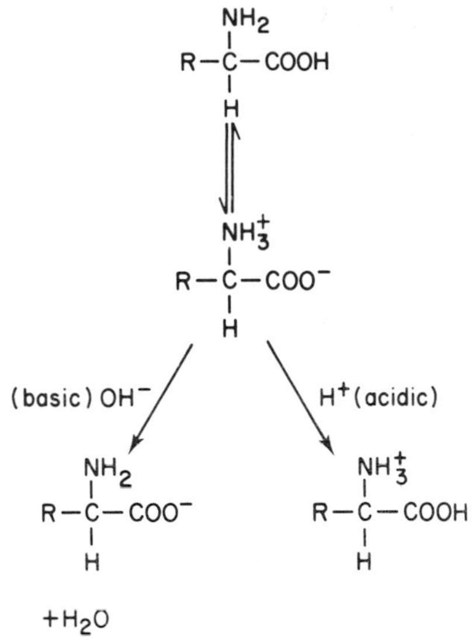

$$NH_2$$
$$R-\overset{\displaystyle |}{\underset{\displaystyle |}{C}}-COOH$$
$$H$$

$$NH_3^+$$
$$R-\overset{\displaystyle |}{\underset{\displaystyle |}{C}}-COO^-$$
$$H$$

(basic) OH^- H^+ (acidic)

$$NH_2$$
$$R-\overset{\displaystyle |}{\underset{\displaystyle |}{C}}-COO^-$$
$$H$$

$$NH_3^+$$
$$R-\overset{\displaystyle |}{\underset{\displaystyle |}{C}}-COOH$$
$$H$$

$+H_2O$

As will be seen the side chains (R) of a number of amino acids also possess amino or carboxylic acid groups which may also develop charges, the polarity of which is dependent on surrounding pH conditions. The charges possessed by amino acids are important in determining their solubility (they are least soluble at their isoelectric point) and in their separation by processes such as ion-exchange chromatography. When combined in protein the charges borne on the individual amino acids play an important role in determining the three-dimensional structure of the protein and in the operation of the active centres of enzymatic proteins.

Amino acids differ from one another in the nature of their side chains which vary considerably in length and complexity. The simplest of the amino acids is glycine where the side chain R is simply a hydrogen atom.

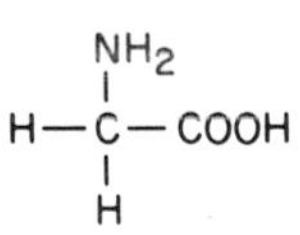

$$NH_2$$
$$H-\overset{\displaystyle |}{\underset{\displaystyle |}{C}}-COOH$$
$$H$$

Glycine

In alanine, R is a single methyl group while valine has a longer, branched side chain.

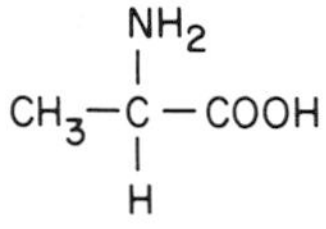

$$NH_2$$
$$CH_3-\overset{\displaystyle |}{\underset{\displaystyle |}{C}}-COOH$$
$$H$$

Alanine

$$NH_2$$
$$CH_3-CH-\overset{\displaystyle |}{\underset{\displaystyle |}{C}}-COOH$$
$$CH_3 \quad H$$

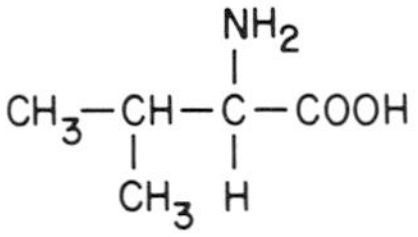

Valine

These side chains may contain additional acidic groups, amino groups, aromatic groups, sulphur groups or hydroxyl groups and provide a convenient means of classifying the amino acids.

4.2 Classification of amino acids

The amino acids can be classified into the following groups.

(a) *Neutral aliphatic amino acids*, for example,

$$
\begin{array}{ccc}
& NH_2 & \\
& | & \\
H-&C-COOH & \\
& | & \\
& H &
\end{array}
\qquad
\begin{array}{ccc}
& H & NH_2 \\
& | & | \\
CH_3-CH_2-&C-C-COOH & \\
& | & | \\
& CH_3 & H
\end{array}
$$

Glycine · Valine

$$
\begin{array}{c}
NH_2 \\
| \\
CH_3-C-COOH \\
| \\
H
\end{array}
\qquad
\begin{array}{c}
\quad\quad\quad NH_2 \\
\quad\quad\quad | \\
CH_3-CH-CH_2-C-COOH \\
| \quad\quad\quad | \\
CH_3 \quad\quad H
\end{array}
$$

Alanine · Leucine

These amino acids have a branched or unbranched simple aliphatic side chain.

(b) *Acidic amino acids*, for example,

$$
\begin{array}{c}
\quad\quad NH_2 \\
\bullet \quad | \\
HOOC-CH_2-C-COOH \\
| \\
H
\end{array}
\qquad
\begin{array}{c}
\quad\quad\quad NH_2 \\
\quad\quad\quad | \\
HOOC-CH_2-CH_2-C-COOH \\
| \\
H
\end{array}
$$

Aspartic acid · Glutamic acid

These two amino acids possess an additional carboxylic acid group in their side chains and therefore have acidic properties. Having two acid groups they are known as dicarboxylic acids. In the ionized form they are referred to as aspartate and glutamate.

(c) *Basic amino acids*, for example,

$$
\begin{array}{c}
\quad\quad\quad\quad\quad\quad NH_2 \\
\quad\quad\quad\quad\quad\quad | \\
H_2N-CH_2-CH_2-CH_2-CH_2-C-COOH \\
| \\
H
\end{array}
\qquad
\begin{array}{c}
\quad\quad\quad\quad NH_2 \\
\quad\quad\quad\quad | \\
HC=C-CH_2-C-COOH \\
| \quad | \quad\quad | \\
N\,\diagdown_{\displaystyle C}\diagup NH \quad H \\
| \\
H
\end{array}
$$

Lysine · Histidine

$$
\begin{array}{c}
\quad\quad\quad\quad\quad\quad\quad\quad NH_2 \\
\quad\quad\quad\quad\quad\quad\quad\quad | \\
H_2N-C-NH-CH_2-CH_2-CH_2-C-COOH \\
\| \quad\quad\quad\quad\quad\quad\quad\quad | \\
NH \quad\quad\quad\quad\quad\quad\quad\quad H
\end{array}
$$

Arginine

These amino acids possess one or more additional amino or imino groups
in their side chains, giving basic properties to the molecules. A shortage of
lysine in the protein of cereal seed (especially maize) is in part responsible
fot the high degre of malnutrition found in areas in which these plants form
the staple diet.

(*d*) *Hydroxy amino acids*, for example,

$$HO-CH_2-\underset{\underset{H}{|}}{\overset{\overset{NH_2}{|}}{C}}-COOH \qquad CH_3-\underset{\underset{HO}{|}}{\overset{\overset{H}{|}}{C}}-\underset{\underset{H}{|}}{\overset{\overset{NH_2}{|}}{C}}-COOH$$

Serine   Threonine

These aliphatic amino acids possess a hydroxyl group in their side chain.

(*e*) *Sulphur amino acids*, for example,

$$CH_3-S-CH_2-CH_2-\underset{\underset{H}{|}}{\overset{\overset{NH_2}{|}}{C}}-COOH \qquad HS-CH_2-\underset{\underset{H}{|}}{\overset{\overset{NH_2}{|}}{C}}-COOH \qquad \begin{matrix} CH_2-S-S-CH_2 \\ H-\underset{\underset{COOH}{|}}{C}-NH_2 \quad H-\underset{\underset{COOH}{|}}{C}-NH_2 \end{matrix}$$

Methionine   Cysteine   Cystine

These amino acids contain sulphur in their side chains. Two molecules of
cysteine can join together via their sulphydryl group to form the 'double'
amino acid cystine. This sulphydryl bridge is very important in holding
together the three dimensional (tertiary) structure of proteins where the
cysteine in one side arm links with that in an adjacent sidearm to form
cystine. The low quantities of the sulphur amino acids present in important
tuber crop plants, such as sweet potato and cassava, are partially responsible for the high degree of malnutrition in parts of the world in which
these plants form the staple diet.

(*f*) *Aromatic amino acids*, for example,

Phenylalanine   Tryptophan

Tyrosine

These amino acids contain aromatic groups in their side chains.
Tryptophan is another amino acid whose low concentration in maize seed is
partially responsible for malnutrition in populations dependent on this
cereal for their basic food supply. The problem is further compounded by

the fact that tryptophan is a precursor of the B vitamin, nicotinic acid (niacin), which forms part of the NAD molecule, a shortage of which causes the disease beri-beri in human beings.

(*g*) *Imino acids*, for example,

$$CH_2-CH_2 \quad\quad HO-CH-CH_2$$
$$CH_2\;CH-COOH \quad\quad CH_2\;CH-COOH$$
$$N \quad\quad\quad\quad\quad\quad N$$
$$H \quad\quad\quad\quad\quad\quad H$$

Proline   Hydroxyproline

Although strictly speaking not an amino acid, proline forms an important constituent of proteins. In a number of plants it is produced in large quantities under conditions of water stress, possibly as an osmoticum to reduce the water potential. Hydroxyproline is an important constituent of cell wall protein.

(*h*) *Amides*, for example,

$$NH_2$$
$$H_2N-C-CH_2-CH_2-C-COOH \quad\quad H_2N-C-CH_2-C-COOH$$
$$O \quad\quad\quad\quad\quad\quad\quad H \quad\quad\quad\quad O \quad\quad\quad H$$

Glutamine   Asparagine

The amides of the two dicarboxylic acids, glutamic and aspartic acids, play important roles in nitrogen assimilation and translocation in plants, but they are also incorporated into the structure of some proteins.

Over 400 plant amino acids have been identified by phytochemical workers (notably Fowden's group at University College, London), but only around 20 of these are found as constituents of protein. These *protein amino acids* are: aspartic acid, glutamic acid (acidic group); asparagine, glutamine (amide group); glycine, alanine, valine, leucine, isoleucine (neutral aliphatic group); lysine, histidine, arginine (basic group): cysteine, methionine (sulphur group); tyrosine, phenylalanine, tryptohan (aromatic group); threonine, serine (hydroxy group); proline (imino group). All of these amino acids can be produced by plants but the enzymes for production of some of them are lacking in animals.

There are eight protein amino acids which cannot be synthesized by human metabolism. These are the '*essential amino acids*', which must be provided in the diet; they are listed in Table 4-1. Histidine is also required in the diet of infants.

The *non-protein amino acids* occur as free compounds in the cells of plants and most appear to be by-products of metabolism with, as yet, no identified function. Some are important as nitrogen transport agents, for example, *o*-acetyl-homoserine in the pea plant, citrulline in the alder.

Table 4-1   The 'essential amino acids' for man.

L-Lysine	L-Valine
L-Tryptophan	L-Methionine
L-Phenylalanine	L-Leucine
L-Threonine	L-Isoleucine

When plant organs are extracted with 80 per cent ethonol the *free amino acids* present in solution in the various compartments of the cells are separated from the *bound amino compounds* which are locked into protein molecules. These free amino acids are throught to exist in cellular 'pools' which have different functions, for example the chloroplastic pool would be associated with amino acid assimilation, the cytosolic pool with the products of amino acid assimilation and protein catabolism, and the vacuolar pool with amino acid storage. The proportion of free to bound amino acids will vary from organ to organ. In actively synthesizing plant leaves, as much as half of the amino nitrogen present may be in the free form, whereas in mature storage organs most of the amino compounds are present in the bound (protein) form. The composition of the free amino acid fractions of plant organs usually differs widely from that of the protein fraction and will vary considerably diurnally as amino acids are introduced into, or are removed from, the different pools. Some amino acids show a very rapid turnover rate (assimilatory pools) whereas others are relatively static (storage pools).

4.3   Biosynthesis of amino acids

The operation of the GS-GOGAT pathway of nitrogen assimilation discussed in Chapter 3 results in the production of glutamic acid which acts as the main vehicle for the introduction of amino nitrogen into plant metabolism. From glutamic acid there are various ways in which amino nitrogen is utilized in the synthesis of the other amino compounds.

4.3.1   Transamination

This is a most important process whereby the amino group of glutamate is used in the formation of other amino acids. In the process of transamination (or aminotransfer as it is often called), the α-amino group of an amino acid replaces the α-keto (2-oxo) group of an α-keto organic acid, the reaction being catalysed by a group of enzymes known as aminotransferases or transaminases. The amino acid donating the amino group is transformed into an α-keto acid. In animal cells, pyridoxal phosphate, derived from vitamin B_6, is an essential coenzyme in the process, although there is some controversy as to whether it is necessary in the equivalent reactions in plants.

Aminotransferases occur in the cytosol, chloroplasts and mitochondria and appear to have fairly wide specificity in that they can each operate on a range of substrates (α-keto acids) with similar structure. Two examples of transamination reactions are shown below.

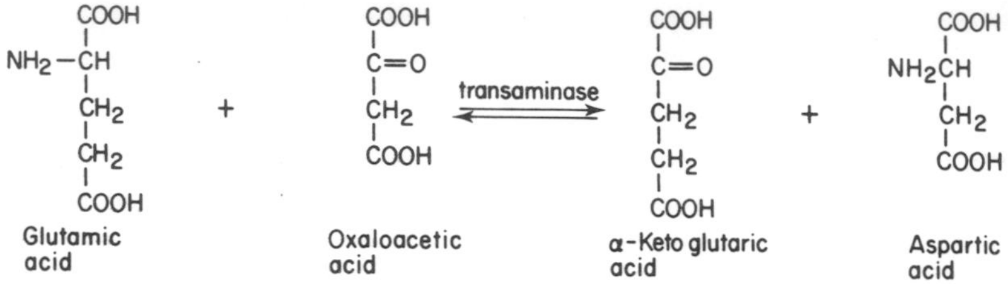

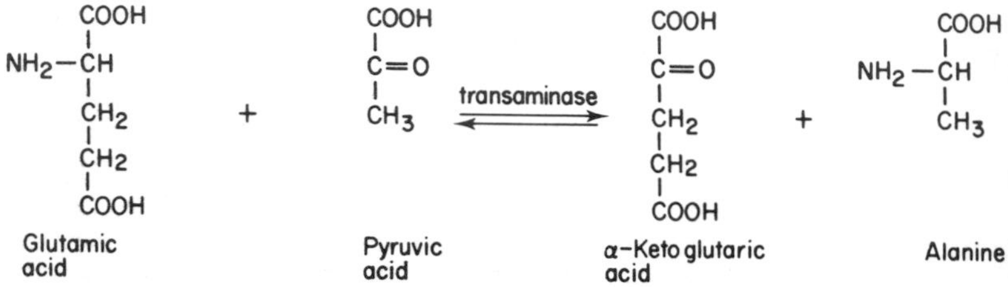

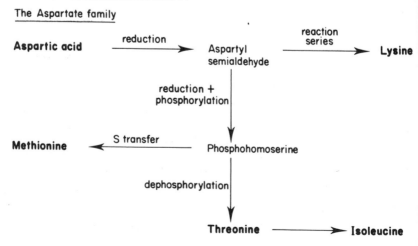

The amino acids produced by transamination from glutamic acid can themselves act as amino donors to α-keto (2-oxo) acids in further transamination reactions.

4.3.2   Elaboration of existing amino acid molecules

Certain amino acids can undergo a series of reactions to produce further amino acids with more complex side-chains. In this way, 'families' of amino acids can be produced from 'family heads'. Examples of 'family head' amino acids are aspartic acid and glutamic acid which can produce 'families' as summarized below.

The Aspartate family

The Glutamate family

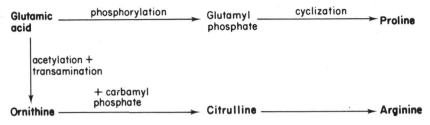

Yet other amino acids are formed by the initial production of their carbon skeletons in a complex series of reactions, before ultimate transamination from glutamic acid or other amino donors (e.g. the production of aromatic amino acids). These are formed by the complex shikimic acid pathway which is summarized as follows:

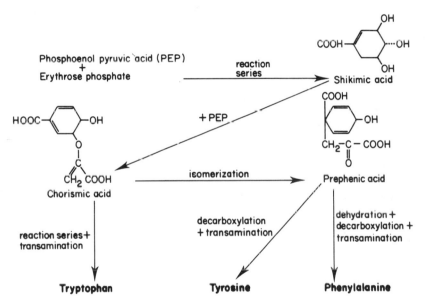

4.4   Origin of the carbon skeletons of amino acids

Although we have been examining in some detail the entry of nitrogen into amino acid assimilation, it must not be forgotten that the bulk of the amino acid molecule is constructed of C and H atoms and that the C-H skeletons must be provided by the plant before amino acid synthesis can take place. These skeletons are provided from the two major sources discussed below.

4.4.1 Photosynthetic origin

If a plant leaf is allowed to photosynthesize for a few seconds in an atmosphere of $^{14}CO_2$ and the amino acids are immediately extracted, separated and their radioactivity determined in a liquid scintillation counter, certain of them show a much higher ^{14}C label than others. This is because the assimilation of their C backbones takes place from compounds derived directly from photosynthetate. The major 'photosynthetic' amino acids are serine, glycine, alanine, aspartic acid and phenylalanine. Serine and glycine are products of glycollic acid produced during the process of photorespiration, which originates in the chloroplast and is continued in the peroxisome and mitochondrion. Serine, in addition, is synthesized by the reduction, dephosphorylation and amination of phosphoglyceric acid, the first stable product of CO_2 fixation in the Calvin cycle. Alanine is synthesized by the amination of pyruvic acid produced from phosphoglyceric acid via phosphoenolpyruvic acid. Aspartic acid arises from the amination of oxaloacetic acid which is produced by the carboxylation of phosphoenolpyruvic acid, and is one of the first major products of photosynthesis in plants which possess the C–4 'aspartate' carbon assimilatory pathway.

The amino acids may be considered an important product of photosynthesis; in some plants undergoing active nitrogen assimilation, as much as 50 per cent of immediate photosynthetic product may be found in this form.

4.4.2   Metabolic origin

All the plant protein amino acids can be produced from intermediates of the major metabolic pathways, e.g. glycolysis, the tricarboxylic acid cycle (Kreb's cycle) and the pentose phosphate pathway, the six most important of these compounds being pyruvic acid, oxaloacetic acid, phosphoglyceric acid, phosphoenolpyruvic acid, 2-oxoglutaric acid, and erythrose-4-phosphate.

A good illustration of the difference in origin of the carbon skeletons of amino acids and their amino groups, reflecting the different amino acid assimilatory mechanisms discussed in Section 4.2 and 4.3, can be obtained by the simultaneous feeding of $K^{15}NO_3$ and $^{14}CO_2$ to the leaves of a plant. The results of such an experiment are shown in Fig. 4-1. Note the prime initial routing of ^{15}N to glutamate and glutamine while newly-assimilated carbon is found primarily in the photosynthetic amino acids serine and aspartic acid.

4.5   Location of amino acid assimilation

Recent evidence has shown that the major amino acid assimilatory sites are the chloroplasts in the leaves (from where the amino acids are apparently able to diffuse into the cytosol) and plastids in the root cell. Other important production sites are mitochondria, peroxisomes and the cytosol itself.

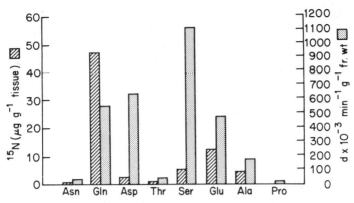

Fig. 4-1 ^{14}C (disintegrations min^{-1} g^{-1} fr. wt.) and ^{15}N (μg g^{-1} fr. wt.) incorporation into the commonest free amino acids of *Datura* leaves fed K ^{15}NO$_3$ and ^{14}CO$_2$. (Ala = alanine; Asp = aspartate; Asn = asparagine; Gln = glutamine; Glu = glutamate; Pro = proline; Ser = serine; Thr = threonine) (From Lewis, O.A.M. (1975). *J. exp. Bot.*, **26**, 361–66.)

4.6 Control of amino acid biosynthesis

The production of amino acids by plant cells appears to be under strict control, the nature of which has not as yet been established. Most experimental evidence points to the control being of the feedback inhibition type, where the end-product of the amino acid synthesis pathway acts as an inhibitor of an enzyme governing one of the reactions early on in the pathway. In this way, if an amino acid is being produced in excess of metabolic requirements and accumulates in the cell, it will repress its own biosynthesis, as for example in the biosynthetic pathway of amino acid F from substance A shown here:

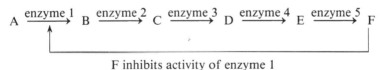

F inhibits activity of enzyme 1

5 Proteins

The important rôle of proteins in biology has been recognized for a long time, and as far back as 1828 the German chemist Mulder coined the class name from the Greek work *'proteos'* meaning 'of prime importance'. Proteins owe their pre-eminent position in Biology to the two main rôles that they play in cells.

(*a*) Many proteins are enzymes responsible for control, through catalysis, of nearly all the reactions that take place in the cell. It is through the production of these enzymatic proteins that the expression of an organism's genes is manifested and the ordered control of its development and metabolism made possible.

(*b*) Other proteins act in a structural rôle where they form important structural units of the cell, for example, membrane and cell wall components, the histone core of chromatin, the mitotic spindle, cellular microtubules and the microtrobecular framework of the protoplasm. The three latter structures are also associated with movement in the cell (e.g. chromosome migration, vesicle migration and cyclosis). In animals this structural rôle is even more prominent, where the production of muscle, skin, connective tissues and materials such as hair and feathers takes place on a large scale.

5.1 Properties of proteins

Proteins are very large molecules, with molecular weights ranging from 6 000 daltons for a small protein to over 2 000 000 daltons for a large one. Because of their size they do not enter into true solution, but form colloidal solutions. Being composed of amino acids, proteins are amphoteric substances and can carry different charges at different pHs (see Section 4.1). Their solubility is lowest at their isoelectric point, the pH which produces equal numbers of positively and negatively charged molecules, thus causing the protein molecules to aggregate together into large masses resulting in precipitation.

Protein solubility is also affected by the concentration of salts in the colloidal solution. Protein sols are hydrophilic, binding a shell of water molecules around themselves and thus preventing coagulation and precipitation. The addition of a salt, such as sulphate to a protein sol, reduces the water available for the formation of the aqueous shell and results in the precipitation of the protein. Different concentrations of salts are required for the 'salting out' of different proteins, thus providing a crude method for the separation of certain proteins.

Proteins are thermolabile substances, that is, their structure is easily damaged by temperatures considerably below that of boiling water. Their themolability is due to the ease with which the bonds responsible for the secondary structure of the protein (see Section 5.2.2) can be dislocated by thermal energy, causing 'denaturation'. The thermolability of enzyme proteins is partially responsible for the narrow range of temperature over which life is possible and why most organisms suffer impaired metabolism at temperatures much over 40°C. It is interesting to note, however, that some organisms can synthesize isoenzymes which are resistant to denaturation (e.g. certain desert plants which can still photosynthesize at 50°C, and some blue-green algae, that are capable of living in hot springs at 70°C).

5.2 Protein structure

Protein molecules are polymers of amino acids (peptides) joined end to end to form long polypeptide chains of varying length and complexity. They are made up of the elements carbon, hydrogen, nitrogen, oxygen and sulphur, the empirical formula of a typical protein, edestin, being $C_{622}H_{1020}N_{193}O_{201}S_4$. The protein molecule can be organized at a number of structural levels, which are referred to as its primary, secondary, tertiary and quaternary structure.

5.2.1 Primary structure

This refers to the basic number, kind and arrangement of amino acids in the polypeptide chain of the protein molecule.

The amino acid content can be determined by heating the protein in 6N hydrochloric acid for 24h at 105°C in a sealed nitrogen atmosphere (to prevent oxidative damage), and separating the products of this hydrolysis by a chromatographic technique such as ion exchange chromatography. The detection and quantitative estimation of the amino acids is carried out by heating them with ninhydrin (triketohydrindene hydrate) in a reducing environment. This reaction produces a purple compound, the optical density of which can be read in an absorptiometer. Cysteine, methionine and tryptophan are wholly or partially destroyed by the hydrolytic procedure and have to be estimated separately.

While estimating the amino acid composition of a protein is relatively simple and rapid using modern techniques, the determination of the actual sequence of the amino acids in the polypeptide chain is a more complex and laborious process, involving the splitting off and determination of each amino acid sequentially from the end of a chain. Consequently, the amino acid sequence of only a few proteins has been determined. The first protein to have its amino acid sequence fully worked out was insulin, a major achievement of the protein biochemist Sanger in 1951. This sequence is shown in Fig. 5-1.

Nearly all proteins have the full complement of protein amino acids (Section 4.2) in their primary structure, but differ from one another in the relative number and sequence of these amino acids in their polypeptide

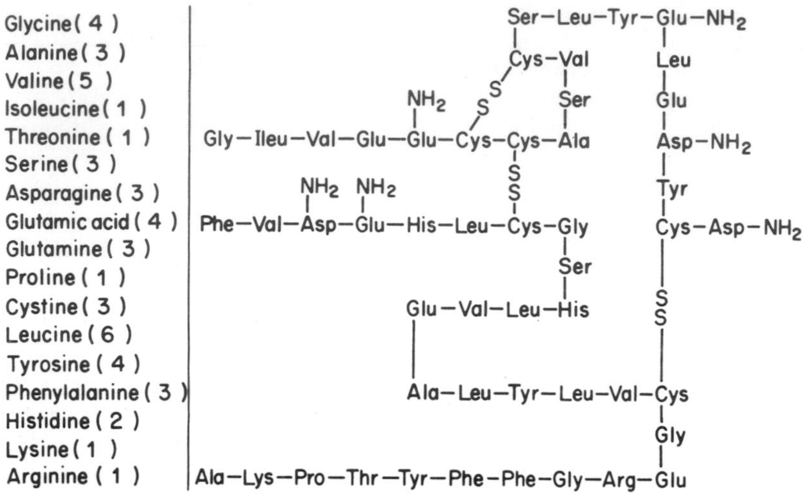

Glycine (4)
Alanine (3)
Valine (5)
Isoleucine (1)
Threonine (1)
Serine (3)
Asparagine (3)
Glutamic acid (4)
Glutamine (3)
Proline (1)
Cystine (3)
Leucine (6)
Tyrosine (4)
Phenylalanine (3)
Histidine (2)
Lysine (1)
Arginine (1)

Ser—Leu—Tyr—Glu—NH₂
          | |
        Cys—Val   Leu
          | |
        S   Glu
  NH₂   S   Ser |
   | S | Asp—NH₂
Gly—Ileu—Val—Glu—Glu—Cys—Cys—Ala |
                     S   Tyr
  NH₂ NH₂   S |
   | | S   Cys—Asp—NH₂
Phe—Val—Asp—Glu—His—Leu—Cys—Gly |
                         Ser |
                          | S
              Glu—Val—Leu—His   S
                          | |
                          | |
              Ala—Leu—Tyr—Leu—Val—Cys
                                     |
                                    Gly
                                     |
Ala—Lys—Pro—Thr—Tyr—Phe—Phe—Gly—Arg—Glu

Fig. 5-1   The amino acid sequence of the protein, insulin.

chains. A small protein such as insulin may have only 50 or so amino acids in its polypeptide chain, whereas a large protein, like glutamate dehydrogenase, may have over 16 000.

When amino acids join together to form chains they do so by forming *peptide bonds*. These bonds arise by the condensation of the carboxyl acid group and the amino acid group of adjacent amino acids in the following manner:

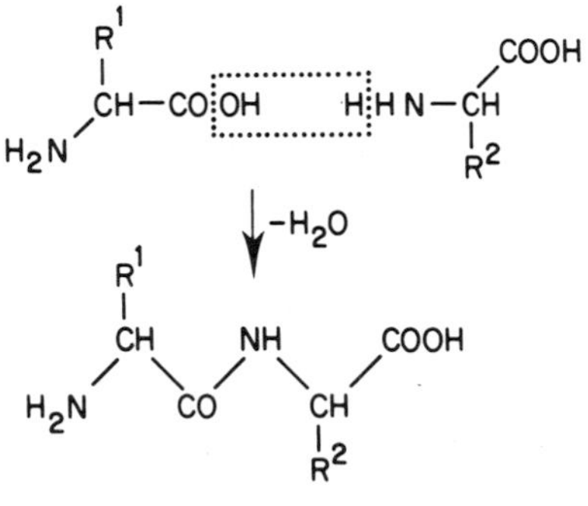

Dipeptides and tripeptides result when two or three amino acids, respectively, join together by peptide bonding, and when a large number condense together a polypeptide results. Note the concertina-like backbone of the polypeptide molecule which contributes to its flexibility.

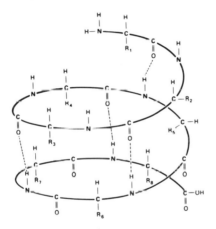

R^1 ... R^3 structural formula

5.2.2 Secondary structure

Weak attractive forces that exist between the carboxyl oxygen of one peptide bond in a linear chain of amino acids and the hydrogen atom linked to the nitrogen of a peptide bond further along the polypeptide chain cause distortions of the protein molecule, producing a secondary structure. X-ray crystallography has shown that this secondary structure can take the form of a spiral where the polypeptide is held as an α-helix with 3.7 amino acids present in each turn of the helix (Fig. 5-2). Many proteins have at least a part of their molecules contorted into this form.

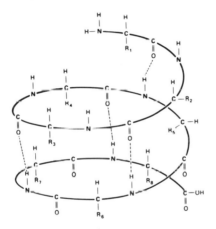

Fig. 5-2   Protein secondary structure : the α-helix. (From Bidwell, R.G.S. (1979). *Plant Physiology*, 2nd ed. Macmillan, New York.)

Another type of secondary structure exists where the weak hydrogen bonds form between the H and the carboxyl O atoms associated with the peptide bonds of parallel polypeptide chains. This is known as the pleated sheet or β-form of secondary structure (Fig. 5-3) and is found in tissue such as muscle protein.

5.2.3 Tertiary structure

The polypeptide backbone may be further convoluted to take on an intricate globular shape. This shape is maintained by disulphide bonds (strong covalent bonds) which form bridges between cysteine molecules in

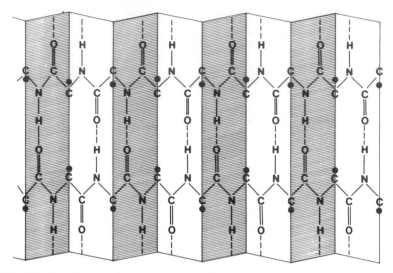

Fig. 5-3   Protein secondary structure : the β-form (pleated sheet). (After Curtis, H. (1979). *Biology*, 3rd ed. Worth Publishers Inc., New York.)

the same polypeptide chain, and also bonds between acidic and basic side chains of various amino acids. (Disulphide bonds may be observed in the structure of insulin, Fig. 5-1, although here they serve to hold two poly-peptide chains together.) Hydrogen bonding and Van der Waal's forces have also been implicated in the formation of protein tertiary structure. The globular shape of a protein exhibiting tertiary structure is shown in Figure 5.4. Note that the secondary α-helix structure is also evident in the arms of the molecule.

The molecular architecture produced by the tertiary structure of a protein is very important in enzyme molecules as it gives rise to crevices into which substrate molecules can fit, thus forming the active centre of the enzyme. When the protein is denatured by heat or large pH changes the secondary and therefore tertiary structure of the enzyme is disrupted and the active centres destroyed, together with the catalytic activity of the enzyme.

5.2.4   Quaternary structure

In a few proteins the complete molecule is composed of a group of poly-peptide chains (each with its own primary, secondary and tertiary structure) which associate together to give the protein its full properties. Such a protein is haemoglobin, composed of four almost identical subunits (Fig. 5-5).

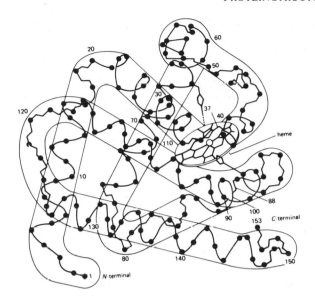

Fig. 5-4 Protein tertiary structure : a molecule of myoglobin (containing a haem group). (From Bidwell, R.G.S. (1979). *Plant Physiology*, 2nd ed. Macmillan, New York.)

Fig. 5-5 Protein quaternary structure. The protein illustrated here has four associated sub-groups. (From Conn, E.E. and Stumpf, P.K. (1966). *Outlines of Biochemistry*. John Wiley & Sons, New York.)

5.3   Conjugate proteins

Proteins may also occur in combination with other molecules, producing conjugate proteins. The most important of these protein combinations are:

(a) *Glycoproteins* These are combinations between carbohydrate and protein molecules and in plants are especially associated with the structure of the cell wall and with the seed reserves of leguminous plants. In animal cells they are particularly important in the formation of the glycocalyx where the carbohydrate portion of the molecule plays a vital role in cell recognition.

(b) *Lipoproteins* In plants, these somewhat loose combinations of proteins and lipid molecules are found primarily in the membranes of the cell and its organelles where they form the basic framework of these important structures.

(c) *Nucleoproteins* The chromatin material of the cell nucleus is composed of a combination of deoxyribonucleic acid and proteins, most of which belongs to the basic histone class. These complex molecules are described in Section 6.4.

(d) *Chlorophyll-protein conjugates* These are found in the photo-pigment systems of chloroplast thylakoids and are of vital importance in the light-harvesting reactions of photosynthesis. The nature of the protein-chlorophyll union effects the light absorption characteristics of the chlorophyll molecule, thus producing chlorophyll *a* species with the different absorption maxima found in photosystems I and II.

(e) *Hemoproteins and flavoproteins* Many important enzymes involved in redox (oxidation-reduction) reactions are conjugate proteins, where heme or flavin groups are attached to protein molecules. These compounds are discussed further in Sections 6.6.2 and 6.3 respectively.

5.4   Separation and classification of proteins

No universally applicable system for the classification of proteins has been worked out and protein taxonomy is still a complex and unresolved branch of science. The classical technique for the separation and identification of proteins based on their solubility in different solvents was developed by Osborne in the 1930s for seed proteins and is still used extensively today in spite of its many drawbacks. This method of classification is not as arbitrary as it may seem, because the solubility of proteins is dependent largely on the type of amino acid they contain and the size of their molecules. The proteins that are separated out by this method are, however, groups of proteins which can be separated into different fractions using more sophisticated means. Details of this classical classification system are given in Table 5.1.

Protein groups that have been isolated by the precipitation methods outlined in Table 5-1 may be further separated into their individual component

Table 5-1   Protein classes as determined by solubility.

Solubility	Protein class	Special features
Soluble in both water and dilute salt solution	Albumins	Includes all enzymes
Soluble in dilute salt solutions, but not in pure water	Globulins	Principal seed protein of oats, peas. Rich in amides, aspartate, glutamate, arginine
Soluble in dilute acids and alkalis but not in water or dilute salt solutions	Glutelins	Large proteins of high molecular weights
Soluble in 80% ethanol, but not in water, dilute salt solutions or dilute acid	Prolamins	High proline and glutamine content. Restricted largely to monocot seeds

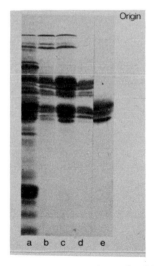

Fig. 5-6   The fractionation of protein by polyacrylamide gel electrophoresis : (a) total proteins of wheat flour; (b) and (c) proteins of the glutenin fraction. (From Payne, P.E. and Rhodes, A.P. In Boulter, D. and Parthier. B. (eds) (1982). Nucleic acids and proteins in plants I, *Encyclopedia of Plant Physiology, (NS) Vol. 14A.* Springer-Verlag, Berlin.)

proteins by more modern analytical techniques. One method is to place the protein fraction on a bed of polyacrylamide or starch gel and apply an electrical current across the bed. The individual proteins will migrate through the bed at a rate dependent on the charge and size of each protein molecule. These charges can be altered by adjusting the pH of the buffer in which the gel is made up. Figure 5-6 shows a separation of the individual proteins of glutenin from wheat using gel electrophoresis.

The method of ion exchange chromatography on diethylaminoethyl (DEAE) cellulose columns (again dependent on the charge of the individual protein molecule), and gel filtration on acrylamide-dextran columns (based on size difference between protein molecules with the acrylamide-dextran beads acting as a sieve) are also used extensively in protein separation.

Apart from the classical protein taxonomic system outlined in Table 5-1, alternative classification systems do exist. Proteins can be classed as 'fibrous' (extended straight molecules, structural in function, e.g. keratin) or 'globular' (polypeptide chain contorted into a globose structure, e.g. 'active' proteins such as enzymes). A similar scheme divides proteins into enzymic, structural (e.g. cell wall, ribosomal or histone proteins) and reserve (seed) categories.

6 Non-protein Nitrogenous Compounds of the Plant

In addition to proteins, the plant produces a large number of nitrogen-containing compounds, many of which play exceedingly important roles in the life processes of the plant. Others may be by-products of nitrogen metabolism and often have no significant function in the biochemical activities of the plant. Many such compounds may nevertheless affect the animal and human organism in various ways, particularly via the nervous system, and are well known as medicines and narcotics.

Some of the more important of these non-protein nitrogenous compounds will be discussed in this chapter.

6.1 The purine and pyrimidine bases

A number of basic compounds are produced by plants, which owe their alkaline properties to nitrogen groups present in their molecules. Amongst the most prominent of these are the purine and pyrimidine bases which are indispensable in numerous biochemical pathways in plants and animals, as well as to the operation of genetic control over all biological processes.

6.1.1 Pyrimidine bases

The best known representatives of this group of bases are cytosine, found in both DNA and RNA, uracil (a component of RNA) and thymine, another pyrimidine base found in DNA. These bases are all derivatives of pyrimidine and have a six-membered ring with nitrogen atoms in the number 1 and 3 positions. Their structures are as follows.

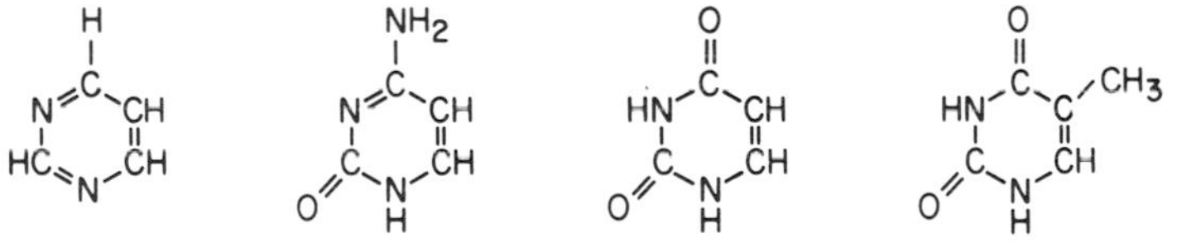

| Pyrimidine | Cytosine | Uracil | Thymine |

The biosynthetic pathway of pyrimidine is complex, the precursors of the ring being carbamyl phosphate and aspartic acid. Carbamyl phosphate itself is formed as follows.

$$\text{Glutamine} + \text{Bicarbonate} + \text{ATP} \longrightarrow \text{Glutamate} + \text{Carbamyl phosphate} + \text{ADP}$$

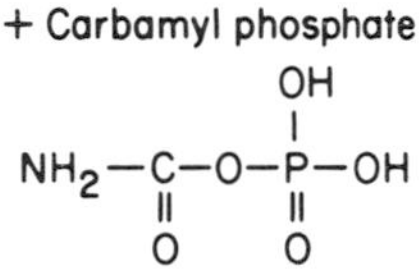

6.1.2   Purine bases

These bases contain a pyrimidine ring which is fused to an imidazole ring, forming the nine-membered purine ring. The major purine bases found in the plant are adenine, guanine (both found in DNA and RNA) and xanthine (the basis of caffeine), which are all derivates of purine.

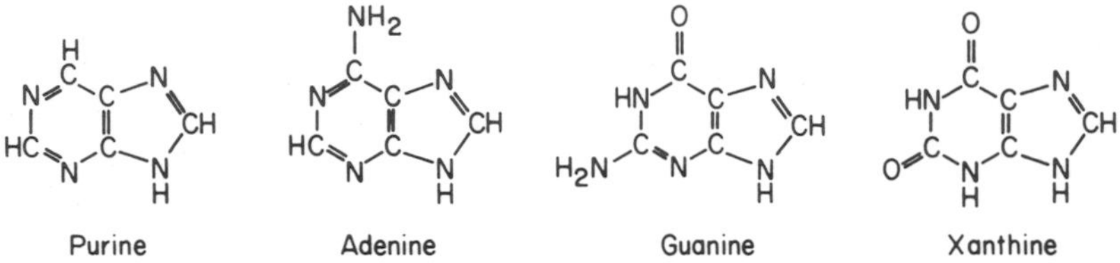

| Purine | Adenine | Guanine | Xanthine |

Like that of pyrimidines the purine biosynthetic pathway is complex, the major precursor compounds being ribose phosphate, asparagine (or glutamine), glycine, aspartic acid and formyl tetrahydrofolate.

6.2   Nucleosides

The purine and pyrimidine bases rarely occur in pure form in the plant, but are usually found combined with a sugar, usually a ribose or deoxy-ribose, to form a nucleoside. One can thus speak of a ribonucleoside (if ribose is the sugar of the nucleoside) or a deoxyribonucleoside (if deoxy-ribose is the sugar of the nucleotide). The structure of a typical ribonucleo-side, adenosine, and a typical deoxyribonucleoside, deoxyadenosine, are shown below.

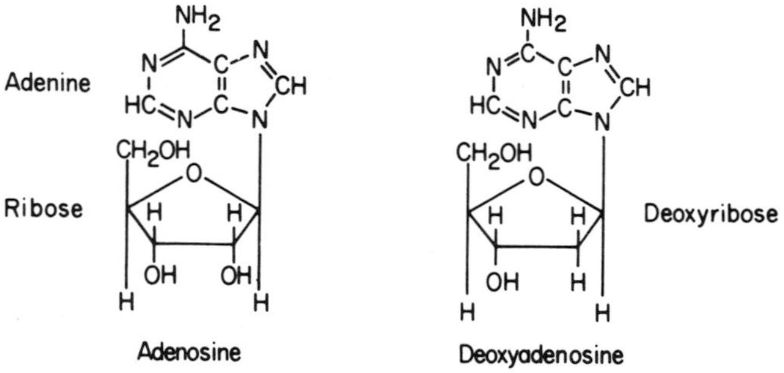

Adenosine   Deoxyadenosine

Note that the structure of the sugar, deoxyribose, is almost identical to that of ribose, the difference being the absence of an oxygen atom on the 2' carbon of the ring (hence the prefix, *deoxy-*).

6.3   Nucleotides

When a phosphate group combines with the sugar residue of a nucleo-side, a nucleotide is produced. This reaction is an esterification and can occur at the 2′, 3′, and 5′ hydroxyl groups of the ribose sugar, the most common reaction being at the 5′ site. A 5′ ribonucleotide or 5′ deoxyribonucleotide results in the latter case.

The 5′ ribonucleotide of adenosine is adenosine monophosphate (AMP), sometimes called adenylic acid or (in the ionized form) adenylate; the 5′ deoxyribonucleotide of adenine is referred to as deoxyadenylate (d AMP). The ribonucleosides guanosine, uridine and cytidine, likewise, give rise to guanylate (GMP), uridylate (UMP) and cytidylate (CMP) respectively.

More than one phosphorylation can take place at the 5′ position on the ribo- and deoxyribonucleotide molecule, resulting in a diphosphate or triphosphate. Thus the formation of adenosine di- or tri-phosphate (ADP, ATP), uridine di- or tri-phoshate (UDP, UTP), and guanosine di- or triphosphate, (GDP, GTP) may occur in cells.

Adenosine monophospate   Adenosine triphosphate

Nucleotides are very important in the overall functioning of the plant for the following reasons:

(a) There are the building blocks from which the nucleic acids are constructed (see Section 6.4).

(b) They act as phosphorylating agents (i.e. phosphate donors) converting organic compounds into their phosphates and thereby enabling them to enter certain reaction sequences otherwise denied to them (respiration etc.), for example,

$$\text{Glucose} + \text{ATP} \xrightarrow{\text{Hexokinase}} \text{Glucose phosphate} + \text{ADP}$$

(c) They can act as energy donors or carriers facilitating many hundreds of reactions in plant and animal cells. Adenosine triphosphate is the best known compound in this connection. Because of the nature of the phosphate bonding which cuts down on the degrees of resonance of the molecule as a whole and juxtaposes negatively charged phosphate radicles which are mutually repelling, far more energy is required to keep the molecule stable. As the 'destabilizing' phosphate radicles are removed, this energy is released and under proper enzyme-controlled conditions, can be used to drive chemical reactions and physical activities in the cell.

$$\text{ATP} + \text{H}_2\text{O} \longrightarrow \text{ADP} + \text{H}_3\text{PO}_4 + 30.5\,\text{kJ}$$

In the reverse reaction, which takes place in processes such as photo-synthesis and oxidative phosphorylation, energy is stored in the ATP molecule.

(*d*) Certain nucleotides, because of their energy content, can act as 'activators' of molecules, enabling them to overcome energy barriers in a number of reactions, for example the role of uridine triphosphate (UTP) in sucrose formation.

$$\text{Glucose phosphate } + \text{ UTP } \longrightarrow \text{ UDP-glucose } + \text{ Pi}$$

$$\text{UDP-glucose } + \text{ fructose } \xrightarrow[\text{synthetase}]{\text{sucrose}} \text{ sucrose } + \text{ UDP}$$

This 'juggling act' makes it energetically possible for the endothermic union of glucose and fructose to take place.

(*e*) Other nucleotides can act as electron donors, acceptors and carriers in a number of redox reactions where they serve as coenzymes to the redox apoenzymes. The best known of these are the 'double' nucleotides, nico-tinamide adenine dinucleotide (NAD) and nicotinamide adenine dinucleotide phosphate (NADP). These large molecules comprise two nucleotides, one containing the base adenine, the other the base nico-tinamide (the functional part of the molecule) and are united via their phosphate groups. NADP contains an additional phosphate group.

Oxidized nicotinamide accepts two electrons, but only one hydrogen atom of the reductant (the other appearing in the solvent). The overall reaction can be written as follows.

Substrate H_2 + NAD$^+$ $\xrightarrow{\text{Apoenzyme}}$ Oxidized substrate + NADH + H$^+$

NADP is the terminal electron acceptor in the light reaction of photo-synthesis, while NAD is probably best known for its redox role in respiration. The flavoproteins, combinations of protein with flavin mono-nucleotide (FMN) or flavin adenine dinucleotide (FAD), are also important redox reagents containing nucleotides. The base flavin (which forms the nucleoside riboflavin) is the reactive part of both molecules.

Oxidized and Reduced flavin

Neither nicotinamide nor riboflavin can be synthesized by human beings and have to be obtained by them in their diet where they form part of the B complex of vitamins. Lack of dietary nicotinamide and riboflavin gives rise to the vitamin deficiency diseases pellagra and beri-beri respectively, both caused by inhibited respiratory function in human cells.

6.4   Nucleic acids

The giant nucleic acids which form the self-replicating genetic control system of all cells are formed from the union of numerous nucleotides, via the phosphate-sugar moieties of each molecule.

6.4.1   Deoxyribonucleic acid (DNA)

This giant molecule which carries the basic genetic information of the cell, is constructed from the deoxyribonucleotides of the four bases adenine, thymine, guanine and cystosine. In each DNA molecule, two chains of these bases form a double helix which interlock with one another and are held together by hydrogen bonds that form between the purine bases (adenine and thymine) of one chain and the pyrimidine bases (guanine and cytosine) of the second chain. In the chromatin material of the cell, this double spiral forms a super-helix which is wrapped around a central core consisting mainly of histone proteins (basic proteins rich in the amino acids arginine and lysine) and a smaller amount of acidic chromatin protein. Five classes of histone proteins group together in clusters of eight, which, together with the DNA wrapped around them, form a nucleosome or 'core unit' (the unit from which chromatin is constructed). This structure is depicted in Fig. 6-1.

The major portion of DNA is found within the cell nucleus, with minor fractions found within mitochondria and chloroplasts.

The sequence of the bases in the DNA molecule forms a code which allows the storage of detailed information on the structure and metabolism of the cell.

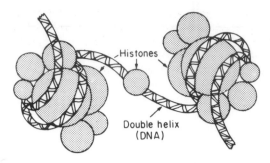

Fig. 6-1 Possible structure of chromatin showing the combination of DNA and histones into nucleosomes or 'core particles'.

6.4.2 Ribonucleic acid (RNA)

This large molecule, formed from the ribonucleotides uracil (not thymine as in DNA), adenine, cytosine and guanine, does not have the double helix structure characteristic of DNA, but the basic structure of the molecule is otherwise the same. Three major types of RNA exist:

(*a*) messenger RNA (mRNA) synthesized on DNA, carries specific instructions for the synthesis of structural and enzymatic proteins to the cytoplasm (it is a short-lived molecule);

(*b*) transfer RNA combines with and activates amino acids, and interacts with ribosomes and messenger RNA during protein synthesis;

(*c*) ribosomal RNA (rRNA) is transcribed from DNA in the nucleolus and is exported to the cytoplasm where it forms part of the structure of ribosomes.

The operation of the RNA genetic code in the control of plant structure and metabolism, is outside the scope of this book, but may be found in many elementary biological textbooks.

6.5 Alkaloids

The alkaloids are another group of nitrogenous bases found in plants; they have occupied the attention of organic chemists for many years because of their interesting ring structures and the pharmaceutical activity which many of them possess.

The term alkaloid (which means 'similar in form to alkali'), refers to a large and heterogenous group of chemicals which usually possess a heterocyclic ring structure containing basic nitrogen. They can range in complexity from simple amines, like ricinine, to the elaborate molecules found in compounds such as strychnine and solanine.

The alkaloids are scattered erratically through almost all plant taxa and can be found in any organ of the plant. An individual alkaloid is, however, usually restricted in distribution to a closely related group of plants.

The plant families which possess the highest number of alkaloid-producing species are Rubiaceae (e.g. coffee), Papaverceae (e.g. poppy), Solanaceae (e.g. tobacco), Leguminosae (e.g. pea) and Apocynaceae (e.g. oleander). Their function in the plant is still enigmatic and there is very little evidence for any plant biochemical or ecological role.

The most popular explanation of their presence in plants is that they act as a deterrent to animal predation. Certain alkaloids are indeed very poisonous to animals but nevertheless plants containing such alkaloids (e.g. the tobacco plant, rich in nicotine) are often heavily parasitized by insects which appear to be immune to this form of toxicity. A nitrogen storage function has been proposed for alkaloids in alkaloid-rich plants, but these compounds are not likely candidates for this role in view of their low nitrogen content. In all probability they are simply by-products of plant nitrogen metabolism which are stored as waste products in view of the lack of plant excretory mechanisms.

In contrast to their apparent lack of function in plants, alkaloids can profoundly affect animal physiology, mainly through the nervous system, and can act as powerful poisons, stimulants, medicines and drugs.

At presents about 3000 alkaloids have been discovered, distributed amongst approximately 4000 plant species (although many more probably exist). As is the case with proteins, their classification is no simple matter. Two major taxonomic systems have been drawn up, one based on the chemical origin of the alkaloids (which we shall follow in this book) and the other based on their ring structure.

Most of the alkaloids are derivatives of the amino acids tyrosine, phenylalanine, tryptophan, ornithine or lysine. Their biosynthesis follows complex pathways which have only recently begun to be understood. They are usually located in cell vacuoles, but in plants like tobacco are found in the cell walls of the leaf and in the organelles of the petiole.

6.5.1 The ornithine alkaloids

These are produced from L-ornithine via a pathway which produces pyrroline as an intermediate. The best known ornithine alkaloid is nicotine, which is formed by a union of pyrroline with nicotinic acid. (Nicotinic acid is derived from a union of aspartic acid with glyceraldehyde phosphate.) A summary of the reaction is as follows.

Ornithine   Pyrroline   Nicotinic acid

Nicotine

Nicotine is obtained from the tobacco plant *Nicotiana rustica*. In animals, it can act as a respiratory stimulant in low concentrations, but in larger concentrations it has the opposite effect and quickly leads to death.

Atropine (DL-hyoscyamine) is another important alkaloid synthesized from ornithine, a well known source of atropine being *Atropa belladonna*, the deadly nightshade. It is used in small concentrations to treat muscle spasms (particularly intestinal) and, until recently, to act as a pupil dilator to facilitate investigations of the eye. In larger doses it acts as a deadly poison, paralysing the central nervous system. The family Solanaceae (*Datura, Scopolia, Hyoscyamus*) is rich in this and closely related alkaloids. (Because of its pupil dilating effect, it was used to enhance the beauty of female eyes by increasing the size of the pupil, hence the specific name, *belladonna*, with presumably disastrous effects on the eyesight of the habitual user.)

6.5.2 The lysine alkaloids

These are produced from L-lysine via a pathway which produces piperidine as an intermediate. Anabasine (from tobacco) and Lupinine (from lupin) are, perhaps, the best known of these alkaloids.

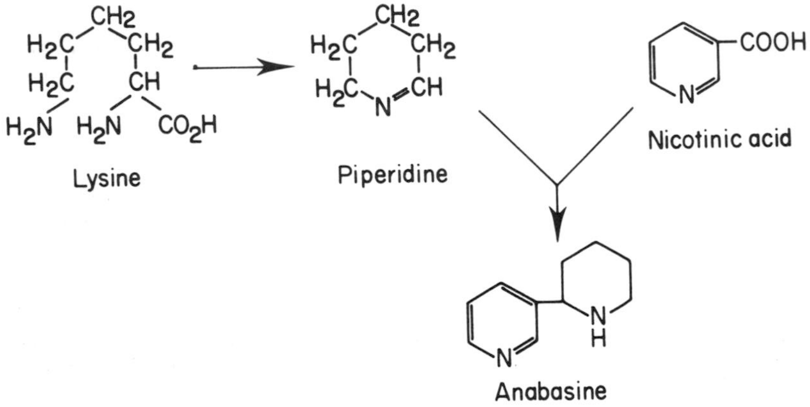

Lysine   Piperidine   Nicotinic acid   Anabasine

6.5.3 The tryptophan alkaloids

These alkaloids characteristically contain an indole nucleus which is derived from the amino acid tryptophan. Various carbon skeletons may be added to this nucleus producing exceedingly complex organic molecules. Well known tryptophan alkaloids are:

(*a*) Lysergic acid, a hallucinogenic drug produced in the ergot fungus (*Claviceps purpurea*) and species of morning glory (*Ipomoea*).

(*b*) Reserpine, from *Rauwolfia serpentina*, which was one of the first 'tranquilizers' discovered. Its leaves had been chewed for many years for this purpose by the inhabitants of parts of India, before the active principle was isolated.

(c) Strychnine, the well known 'detective-book' poison, can be used as a respiratory stimulant in small doses. In larger doses, however, it becomes a powerful toxicant causing prolonged muscular spasm leading to a painful death. It is found in plants of the genus *Strychnos*, the endosperm of *Strychnos nux-vomica* being one of the richest sources of strychnine.

(d) Quinine, so well known in the treatment of malaria, is produced by species of *Cinchona* (Rubiaceae). Although a derivative of tryptophan, the structure of this amino acid has been changed into the quinoline form, where the 5-membered ring of indole has been converted to a nitrogen-containing 6-membered ring. One can thus speak of indole alkaloids (e.g. strychnine) or quinoline alkaloids (e.g. quinine).

| Tryptophan | Strychnine (indole alkaloid) | Quinine (quinoline alkaloid) |

The families Apocynaceae, Loganiaceae, Rubiaceae and Euphorbiaceae are known particularly for their possession of species rich in complex tryptophan alkaloids.

6.5.4   The tyrosine and phenylalanine alkaloids

A number of alkaloids are synthesized from tyrosine and phenylalanine alone, or in combination with other amino acids. Well known examples of this group are as follows.

(a) Cocaine is derived from ornithine and phenylalanine, so could also be correctly placed among the ornithine alkaloids. Its major source is the genus *Coca* and it is used extensively in medicine as a local anaesthetic. Unfortunately it is also abused as one of the most dangerous habit-forming hallucinogenic drugs.

(b) Morphine ('morphia') and codeine are structurally similar alkaloids, derived from the union of two molecules of dehydroxyphenylalanine (DOPA) resulting in a benzylisoquinoline structure. DOPA itself is an oxidation product of tyrosine. The morphine-related alkaloids are often referred to as the benzylisoquinoline alkaloids. Morphine is the most effective pain-killer known to man and is used as such in medicine, although usually only in cases of intractable pain or terminal illnesses, as it is an extremely habit-forming drug. Codeine is a milder analgesic, but appears not to suffer from the fatal habituating effect of morphine. Heroin (diacetylmorphine), manufactured from morphine, is more popular than morphine as a hallucinogen because of the more intense 'euphoria' it produces. It is even more addictive than morphine. The source of the morphine alkaloids is the latex of the unripe capsules of the poppy plant, *Papaver somniferum*. Opium, the unrefined extract of the latex, contains a number of alkaloids including morphine, narcotine, codeine, papaverine and thebaine.

Tyrosine   DOPA   Morphine

6.5.5 Purine alkaloids

In addition to amino acids, the purine ring can also act as a major precursor of alkaloid formation. The best known of the purine alkaloids is caffeine, the mentally-stimulating drug found in coffee (*Coffea arabica*), tea (*Camellia sinensis*) and cocoa (*Theobroma cacao*). The purine base, xanthine, is the basis of the structure of caffeine and the allied alkaloid, theobromine (found in the cocoa plant).

Xanthine   Caffeine

6.6 Porphyrins

These are large, tetrapyrrolic molecules, the best known plant representatives of which are chlorophyll and the cytochromes.

6.6.1 Chlorophyll

In the green plant one of the most important and plentiful compounds is chlorophyll. Chlorophyll belongs to a class of a compounds known as the porphyrins. Each prophyrin possesses a tetrapyrrole ring, that is, it is formed from four nitrogen-containing pyrrole rings linked together by methylene groups to form a ring system. In the case of chlorophyll, a magnesium ion is chelated to the nitrogen atoms of the pyrrole rings, taking up a position in the centre of the molecule.

Pyrrole ring

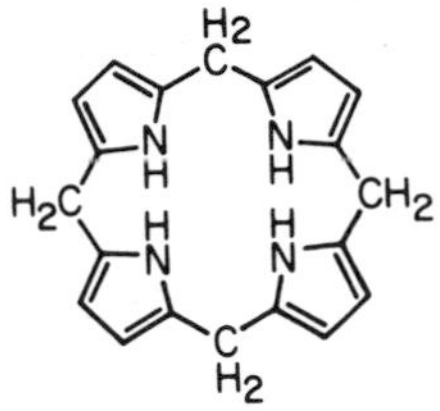

Tetrapyrrole ring

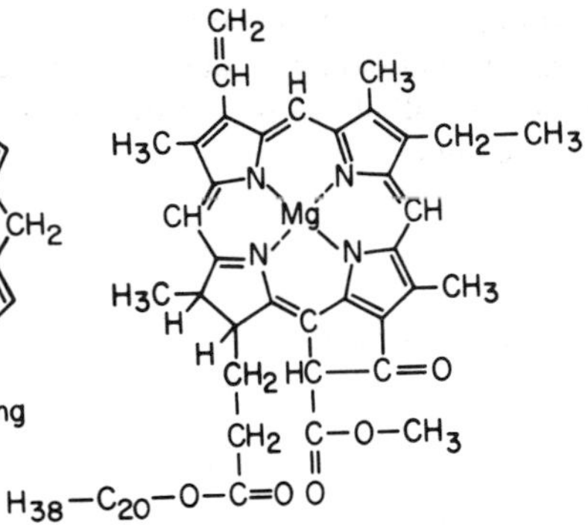

Chlorophyll *a*

The synthesis of the chorophyll molecule is complex and not all the steps have as yet been elucidated with certainty. It is basically as follows:

(*a*) glumatic acid is converted via glutamate semialdehyde to δ-amino-laevulinic acid (ALA);

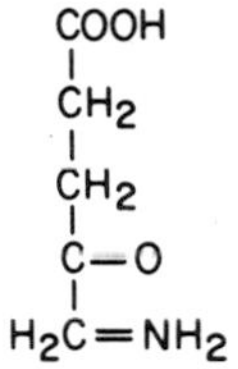

$$COOH$$
$$|$$
$$CH_2$$
$$|$$
$$CH_2$$
$$|$$
$$C-O$$
$$|$$
$$H_2C=NH_2$$

δ-aminolaevulinic acid

(*b*) two molecules of aminolaevulinic acid condense to form a pyrrole ring known as porphobilinogen;

(*c*) four porphobilinogen molecules become linked in a reaction producing a tetrapyrrole, *Uroporphyrinogen*;

(*d*) *uroporphyrinogen* is converted in a series of oxidation and decarboxylation reactions to *protoporphyrin* at which stage Mg^{2+} becomes enzymatically inserted into the molecule;

(*e*) a further series of reactions results in *protochlorophyllide a* and this molecule, on the attachment of a phytyl chain, becomes *chlorophyll a*.

Linear tetrapyrroles may be found in plants; these structures form the basis of molecules like the flowering regulator, phytochrome, and the algal phycobiliproteins (phycocyanin and phycoerythrin pigments).

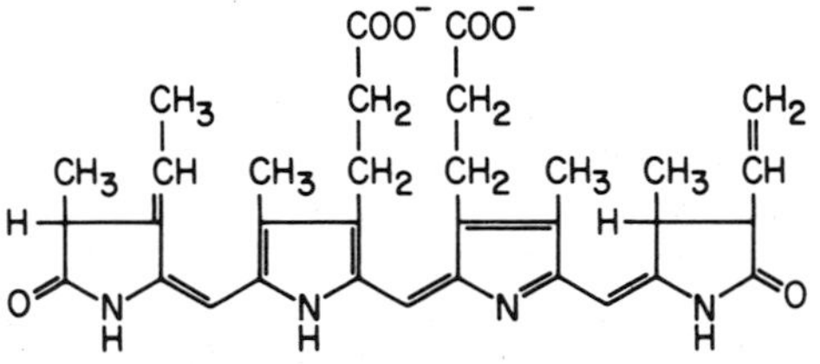

Phytochrome

6.6.2 Cytochromes

Another group of important porphyrin-containing compounds is the cytochromes, which are vital in many cell redox processes, particularly oxidative phosphorylation and photophosphorylation, as electron acceptors and donors. The working cytochrome molecule is a combination of a protein and a molecule of heme. Heme has a structure very similar to chlorophyll; it is a tetrapyrrolic compound, but in place of magnesium, iron is chelated into the central part of the molecule. The iron can be in the Fe^{2+} or Fe^{3+} state, depending on whether or not the molecule is reduced. It is thus the heme group which is responsible for the redox capabilities of cytochrome. The operation of the cytochrome redox system can be considered enzymatic, with the protein moiety acting as the apoenzyme and the heme as the prosthetic group. The biosynthesis of the cytochromes is similar to that outlined for chlorophyll, with iron protoporphyrin being produced in place of magnesium protoprophyrin.

The cytochromes are divided into a number of classes (a, b, c, d, e and f), depending on the structure of the heme moiety and its bonding to the protein moiety. Varieties of each class are known and some have been given subscripts to identify them (e.g. cytochrome a_1, a_2, etc.).

Cytochromes are usually identified from their absorption spectra and it is usual to designate a newly discovered cytochrome according to the wavelength of its absorption peak, which lies between 545 and 600 nm for the reduced molecule, (e.g. Cytochrome a_{572}). Cytochrome complexes can arise, such as the 'cytochrome oxidase' of yeast which consists of a group of protein and heme molecules.

Protoheme (prosthetic group of the *b* cytochrome)

6.7 Phytohormones

Two of the major plant growth hormones, the auxins and cytokinins, are also important nitrogen-containing substances, even through they occur only in minute amounts in plants.

The most abundant naturally occurring auxin is indole-3-acetic acid which is a close derivative of the amino acid L-tryptophan.

L-tryptophan   Indole 3-acetic acid

Most of the cytokinins, of which there is an assortment, are derived from the base adenine, but the biosynthetic pathway is not well understood. The flowering regulator, phytochrome, is discussed in Section 6.6.1.

Zeatin (a cytokinin)

In the last two chapters we have discussed the more important nitrogen-containing compounds found in plants. Other nitrogenous compounds do exist, for example, the amines (decarboxylation products of amino acids); cyanogenic glycosides such as amygdalin of almond nuts which liberates HCN on chewing, and the mustard oil glucosides (derivatives of valine and phenylalanine) which give characteristic odours to the leaves of the family Cruciferae. Space precludes their further discussion in this book.

7 Transport of Nitrogen in Plants

The chief pathways for the long distance translocation of nitrogeous compounds in the plant are the xylem and phloem with their specialized transport cells. The xylem is the major route for the transport of both inorganic nitrogen (especially nitrate) and the organic nitrogen products of root metabolism, from root to shoot. This takes place via the transpiration stream. The phoem, on the other hand, is responsible for the translocation of nitrogenous compounds manufacture in the leaf, not only to the fruit and growing regions of the plant (which have a weak transpirational pull), but also back to the root. The plant's 'plumbing plan' with respect to nitrogen translocation is shown in Fig. 7-1.

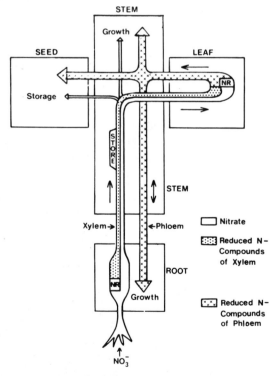

Fig. 7-1 The plant plumbing plan showing the transport of nitrogenous compounds. (NR = nitrate reductase).

7.1 Translocation of nitrogen through the xylem

Xylem sap may be examined by collecting the 'bleeding sap' exuding from the stumps of decapitated plants (Fig. 7-2) or the fluid extracted from the shoots of plants by placing their severed ends under vacuum and slowly cutting the shoot back. Phloem sap is not extracted by these methods, as in most plants the phloem seals itself off by rapid deposition of callose on its sieve plates almost immediately after wounding and contact with air. In spite of the possibility of contamination of the extracted xylem sap with the contents of cells damaged by decapitation, it is widely accepted that analysis of this sap provides a reasonably accurate picture of the xylem transport of nitrogen. The rate of delivery of the nitrogenous solute to the shoot may be approximated once the transpiration rate of the plant is known.

Xylem sap is acidic with a pH of between 5.4 and 6.5 (phloem sap is alkaline). Nitrogenous compounds form the major dry matter component of xylem sap with nitrogen levels in the range of 0.01–0.21 per cent (w/v). The C:N weight ratio of the sap lies within the range of 1.5 to 6 (compared with the 15 to 200 for pholem sap) indicating the dominance of nitrogen-containing compounds in xylem transport.

7.1.1 Compounds of N transport in the xylem

The nature of the compounds which the root of individual species exports to the shoot via the xylem, depends largely on the rôle which that root plays in overall nitrogen assimilation. In some species with relatively weak nitrate reductase activity in their roots, (and where the plant's major nitrogen

Fig. 7-2   Collecting xylem sap from decapitated barley plants.

source is nitrate), over 95 per cent of the xylem nitrogen may consist of free nitrate, for example, *Xanthium* (cocklebur), *Cucumis* (cucumber) and *Gossypium* (cotton). The roots of other species have high nitrate reductase activity, and even under conditions of high nitrate feeding levels show little or no nitrate transfer from root to shoot via the xylem. The absorbed nitrate is almost entirely assimilated in the root and nitrogen is loaded onto the xylem for export to the shoot in fully reduced organic form. Examples of such plants are *Raphanus* (radish), *Lupinus* (lupin) and *Pyrus* (apple). Most herbaceous and woody plants, however, have active nitrate reductase systems in both shoot and root, and both free nitrate and organic nitrogen compounds may be found equally represented in the xylem sap. A spectrum of xylem sap composition from the various plant categories we have discussed is shown in Fig. 7-3.

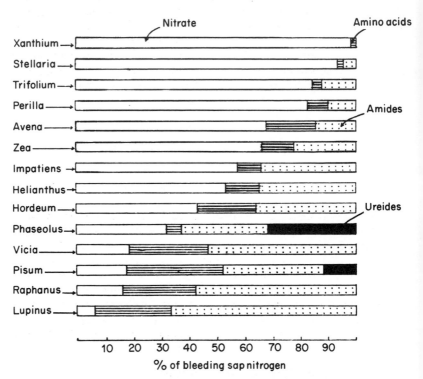

Fig. 7-3   Spectra of nitrogenous compounds found in the xylem ('bleeding') sap of various plants. (From Pate, J.S. (1973). *Soil Biol. Biochem*, **5**, 109–19.)

Plants fed with ammonium carry out the assimilation of this ion almost exclusively in their roots. The xylem stream of such plants carries relatively little ammonium, the major source of nitrogen supply to the leaves being in combined organic form, usually an amide (see Chapter 2).

The main organic compounds exported from root to shoot via the xylem stream are highly characteristic of the plant species. In many plants the nitrogen-rich amides, glutamine and asparagine, are the major nitrogen-containing solutes (e.g. in *Triticum, Hordeum, Secale, Zea, Datura, Solanum* and *Helianthus*). In apple, the bulk of the nitrogen carried in the xylem is in the form of asparagine and arginine. In some plants, non-protein amino acids may be among the major translocated nitrogen compounds, for example, homoserine (*Pisum*) γ-methylene glutamine (*Arachis*), citrulline (*Alnus*); in others, alkaloids can serve this function.

In nodulated legumes the xylem stream is particularly rich in nitrogenous compounds exported from the nodules, where they are produced in response to the activity of the nitrogen-fixing bacteria contained in them. In certain tropical legumes (e.g. cowpea and soya bean), ureides are the main nitrogen translocation compounds, whereas in many temperate legumes (peas, beans, lupins), the amides (especially asparagine) and certain amino acids are the major nitrogen-containing solutes of the xylem sap.

The ureides, synthesized from purines, are nitrogen rich molecules which, with their low C:N ratios, are ideal for nitrogen transport purposes. The two major ureides are allantoic acid and allantoin (both with a 4N:4C ratio) which have the following formulae:

Allantoin   Allantoic acid

A major disadvantage of the ureides as vehicles of nitrogen transport, is their very low solubility compared to that of other translocated nitrogen molecules. Thus most ureide-transporting plants load both allantoin and allantoic acid simultaneously into their xylem stream, to allow adequate quantities of nitrogen to reach the shoot from the root under the constraints imposed by the respective low solubilities of these compounds.

Their low solubilities do, however, allow certain plants to use the ureides as nitrogen storage compounds (e.g. *Acer* spp., *Protea* spp.), although frequently for short periods only, for example, the storage of ureides in developing seeds and seed pods prior to fruit maturation.

7.1.2   Variation in N contents of xylem fluid

The rates of loading nitrogenous compounds from a root onto xylem sap is by no means constant and follows a circadian rhythm in a number of plants, with a maximum at or near midday and a minimum at midnight. The concentration of N compounds in the xylem sap is also dependent on the transpiration rate of the plant at that particular time. If the transpiration rate is fast with a rapid passage of water from root to shoot, the nitrogen concentration of the xylem sap will be relatively low compared to that of the same plant experiencing a slow transpiration rate. Thus, a reasonably constant rate of nitrogen delivery to the shoot is ensured, which is to a certain extent independent of transpiration rate.

Analyses of root bleeding-sap have also shown that the concentration of nitrogenous solutes in the xylem changes in a predictable fashion during development of the plant. During the seedling stages the levels of these solutes are low, but once uptake of nitrogen from the soil commences, their concentration increases rapidly and remains high until after flowering. Hereafter the assimilation and export of nitrogen by roots declines radically.

It must be appreciated, therefore, that the nitrogen composition and content of the xylem sap of a plant varies widely even during the same day, owing to a number of factors, and a wide range of samples should therefore be analysed before definite conclusions are reached regarding the transport of nitrogen through the xylem of a plant species.

7.2   Translocation of nitrogen through the phloem

Phloem sap contains much higher concentrations of nitrogenous compounds than xylem sap, with nitrogen levels sometimes reaching 4 per cent (w/v) of the fluid. The major content of the phloem sap is, however, carbohydrate (usually sucrose): consequently the C:N ratio of phloem sap is much higher than that of xylem sap and ranges from 15 to 200 (see Section 7.1).

The extraction of phloem sap from plants is far more difficult than the extraction of the xylem sap, because of the 'self-sealing' nature of the sieve tubes of most plants once they have been damaged. Various techniques do exist, however, and these are described below.

(a) In a number of plants which are subject to attack by aphids, it is possible to make use of the fact that the aphid stylets penetrate only the phloem cells of the plant and thus extract pure pholem sap for consumption. If the aphids are carefully anaesthetized and their bodies cut away from their stylets, it is possible to collect the phloem sap exuding from the stylets under the positive pressure which exists in the phloem (see Figs 7-4 and 7-5). The sap thus collected is considered to be very pure, but the method is not completely satisfactory as very little sap can be collected from each stylet and the aphids feed at somewhat restricted locations on a plant.

(b) An alternative method can be used in a large number of woody plants and some non-woody plants (e.g. peas, lupins, castor-oil, cucumber and beet), where the phloem is slow to seal after wounding. The sap exuding from shallow incisions in the bark of the woody plants or the cut surfaces of fruit pods, petioles and inflorescence stalks of the slow-sealing herbacous plants may be collected in sufficient quantity for analysis. There is, however, the danger of contamination from the contents of xylem and of cells which have been damaged in the preparation of the specimens. It has even been possible to collect phloem sap from plants with quick-sealing sieve tubes by immersing the severed ends in a solution of chelating agent (EDTA, i.e. ethylenediaminetetraacetic acid), immediately after cutting. This treatment greatly retards the sealing of the sieve tubes, but does pose problems in the eventual separation of the exuded pholem solutes from the chelating solution.

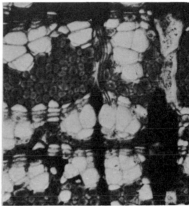

Fig. 7-4 (Left) A droplet of phloem sap exuding from an aphid. (Right) Micrograph of the tip of an aphid stylet penetrating a sieve tube. (From Curtis, H. (1979). *Biology*, 3rd ed. Worth Publishers Inc., New York.)

Fig. 7-5   A drop of phloem sap exuding from the cut tip of a lupin pod. (Pate J.S. and Kaiser J.J.)

Unlike the xylem sap, phloem sap is alkaline in reaction (pH 7.4–8.6). The major nitrogenous compounds transported by the pholem have been found to be usually identical to those transported by the xylem in any one species, with the exception of nitrate which is found only in trace quantities, if at all, in the phloem. Ammonium is also absent from the phloem sap, or present at only very low concentration. A comparison of the nitrogenous contents of xylem and phloem sap of *Brassica* is shown in Table 7-1.

Table 7-1 Comparison of the composition of phloem and xylem sap bleeding from cut young flower stalks of purple sprouting broccoli (variety of *Brassica oleracea*). (From Pate, J.S. (1973). *Soil Biol. Biochem.*, **5**, 109–19.)

Amino compounds (parts $N/10^6$ sap. vol. (w/v))	Phloem sap	Xylem sap
Aspartic acid	29.1	13.1
Asparagine	T	T
Glutamine	T	317.6
Threonine	34.6	T
Serine	70.2	T
Glutamic acid	243.1	73.3
Proline	319.6	0
Glycine	15.7	T
Alanine	48.8	5.5
Valine	39.5	T
Isoleucine	21.0	T
Leucine	32.5	T
Tyrosine	T	0
Phenylalanine	T	0
γ-amino butyric acid	T	17.1
Lysine	51.0	T
Histidine	9.3	T
Arginine	99.8	T
Nitrate (parts $N/10^6$ (w/v))	T	117.7
Sucrose (per cent w/v)	4.3	0

T = trace

7.3 The fate of xylem-borne nitrogen

A number of studies have demonstrated that the largest proportion of the solutes carried in the xylem by the transpiration stream, ends up in the leaves. Analyses of bleeding sap at different levels of a stem show that much of the nitrogen is abstracted for immediate use or storage by the stem itself. The feeding of ^{14}C and ^{15}N labelled amino compounds to detached shoots of *Lupinus albus* by Pate and his co-workers has shown selectivity by the stem in the absorption of these compounds. ^{14}C – Arginine, for example, was absorbed so effectively in a 6h feeding experiment that only 30 per cent of the ^{14}C fed to the plant reached the leaf, whereas 80–90 per cent of the ^{14}C applied as aspartic and glutamic acids reached this organ. There is also much evidence to indicate that xylem to phloem transfer of nitrogenous compounds via transfer cells takes place in the stem, and the significance of this will be discussed in Section 7.5.

When the nitrogenous solutes reach the leaf, they enter leaf metabolism, and compounds such as glutamic acid and glutamine rapidly donate their amino and amide groups for the synthesis of a variety of amino acids in the chloroplasts. These newly formed amino acids, together with others delivered by the xylem stream, may be used to form the proteins of the leaf (particularly the CO_2 trapping enzyme, ribulose-bisphosphate carboxylase, which is the most abundant protein in the world) or may be exported via the phloem to growing or storage regions of the plant. Nitrate must first be reduced to ammonium by the nitrate-nitrite reduction system before it can enter leaf nitrogen assimialtion. In some plants (particularly members of the Amaranthaceae, Chenopodiaceae, Compositae (Asteraceae), Cruciferae and Solanaceae), nitrate may be used as a nitrogen storage compound and can build up to high concentrations (sometimes as much as 5 per cent of tissue dry weight) in leaves and other regions of the plant.

In many plants, xylem-delivered asparagine is metabolized little by leaves and is apparently loaded directly onto the phloem for export to fruit and growing regions of the plant.

7.4 The fate of phloem-borne nitrogen

Phloem-borne nitrogen is the major source of supply of this element to regions of the plant provided with limited transpirational attraction for xylem sap, for example, developing fruit, young expanding leaves, growing regions of the plant, and the root itself. Investigations using ^{15}N indicate that nitrogenous compounds arriving at their delivery sites via the phloem are metabolized and their nitrogen redistributed through a range of amino acids for use in protein building. Glutamine is one of the commonest nitrogenous substances delivered by the phloem; Beevers has shown that in plants such as maize, the enzyme glutamine synthase is active in the fruit, thus enabling the nitrogen of glutamine to become involved in fruit amino acid synthesis. In the developing seeds of plants such as the pea, the presence of the enzyme asparaginase has been demonstrated by Ireland and

Joy; the enzyme would be responsible for the utilization of phloem-delivered asparagine in protein synthesis. In soya beans, Thomas and Schrader have shown the presence of allantoinase activity in the developing fruit. This enzyme facilitates the entry of the nitrogen of the ureide allantoin into fruit amino acid assimilation. It should be noted that rather little is known about the assimilation of xylem-bound ureides in shoot tissue, although the enzymes allantoinase, allantoicase and urease are known to be involved, probably as follows:

$$\text{Allantoin} \xrightarrow[\text{allantoinase}]{\overset{\text{hydrolysis}}{\text{by}}} \text{Allantoic acid} \xrightarrow[\text{allantoicase}]{\overset{\text{hydrolysis}}{\text{by}}} \text{Glyoxylic acid} + \text{Urea}$$

$$\text{Urea} \xrightarrow{\text{urease}} \begin{array}{c} \text{Ammonia} \\ + \\ CO_2 \end{array}$$

A number of transaminating enzymes have also been shown to be present in plant seeds.

In spite of the fact that all the nitrogen absorbed by the plant must pass through the root (apart from that originating from foliar sprays), the root itself is often a strong sink for phloem-borne nitrogenous compounds exported by the leaf. This is particularly true of those plants with low nitrate reductase activity in their roots. Thus, a considerable amount of cycling of nitrogen takes place between root and shoot, sometimes making it difficult to assess the relative activities of root and shoot in nitrogen assimilation.

7.5   Xylem-phloem interchange

Phloem analyses of the white lupin (*Lupinus albus*) made by Pate and his co-workers have revealed that the sap which passes from the leaves is, at all times of growth, much less concentrated in amides and certain amino acids (in relation to sucrose) than is the sap arriving at the fruit or growing points. Thus, the phloem sap leaving the leaves contains *less* nitrogen than that arriving at the sinks. The feeding of ^{14}C and ^{15}N amino compounds to the xylem stream, and their subsequent recovery from the phloem as they pass up the stem, has demonstrated that relatively large-scale transfer of N-compounds takes place from xylem to phloem. In *Lupinus* it is asparagine, glutamine, valine, serine and lysine which are primarily involved in the transfer and are therefore responsible for the marked lowering of the C:N ratio of the phloem sap as it passes from leaves to the various sinks (fruit, roots, growing points). This transfer phenomenon (probably mediated by transfer cells) is considered to be a mechanism for improving the nitrogen supply to these active protein-producing regions of the plant, which are denied direct access to the incoming nitrogen supply of the xylem. This results from their lack of transpirational surface so necessary for the attraction of xylem-borne solute.

Also responsible for the decreasing C:N ratio of the phloem as it passes through stem or petiole tissue, is the loading onto the phloem sap of stored nitrogenous compounds previously abstracted from the xylem by the stem and petiole. Nitrogenous materials released as a result of the senescence of these organs (or leaves associated with them) will also decrease the phloem C:N ratio.

8 Nitrogen Metabolism in the Seed

8.1 Seed formation

One of the major reasons for the success of the Angiosperm plant has been its development of the seed habit which has vastly improved the chances of survival of the plant's progeny by providing protection and nourishment at this critical and vulnerable period of the life cycle. An important feature of seed formation is therefore the laying down of adequate resources of nitrogenous and carbohydrate or lipoid materials to sustain the growing embryo after seed germination, until such time as its own photosynthetic processes have become sufficiently developed to take over this function.

Protein is the major form in which nitrogen is stored in seeds and its content ranges from the 8–10 per cent found in cereal seeds to the 20–30 per cent found in legumes.

As we have seen in Chapter 7 the phloem is the main pathway for nitrogen delivery to the seed. (Pate has estimated that in the white lupin, 89% of the seed's nitrogen arrives via this pathway.) These translocating compounds have a number of sources. They may originate in the leaf as a result of that organ's nitrogen assimilating activity, or the hydrolysis of its protein following the onset of senescence. In some annuals, this latter process may be the major source of the seed's nitrogen supply. Other sources are the nitrogenous reserves of stem and petiole, or the root itself which may supply the seed directly with compounds such as asparagine and various amino acids by xylem to phloem transfer (see Section 7.5). In plants with photosynthetic pericarps (e.g. the pods of peas and beans), these can be another important source of amino compounds for the developing seed. In the seed, the nitrogen of these translocating compounds appears to be redistributed among the full spectrum of amino acids, which are then built into the reserve proteins of the seed.

The main protein storage compartments of the seed are either the endosperm or the cotyledon(s), depending on the plant family. Cereals, for instance, use the endosperm for seed protein storage, whereas in legumes the bulk of the storage protein is found in the cotyledons.

8.1.1 Nature of storage proteins

The nature of the reserve proteins also differs from plant to plant (see Table 8-1). In general, cereals usually store prolamins (40–60%) and glutelins (20–40%), while dicotyledons have globulin as their main storage protein. A notable exception is the oat plant with 80 per cent of its seed protein reserve in the form of globulin.

Table 8-1   The protein content of various seeds. (From Beevers, L. (1976).
Nitrogen Metabolism in Plants. Edward Arnold.)

| | Fractions as per cent of total protein | | | |
	Albumin	Globulin	Glutelin	Prolamin
Wheat (*Triticum aestivum*)	5	10	40	45
Maize (*Zea mays* IND 260)	14	–	31	48
Maize (*Zea mays* Opaque 2)	25	–	39	24
Pea (*Pisum sativum*)	40	60	–	–
Oat (*Avena sativa*)	Trace	80	5	15
Squash (*Cucurbita pepo*)	Very little	92	Small amount	Very little
Rice (*Oryza sativa*)	5	10	80	5

The fact that different seeds possess different ratios of the major protein types (see Section 5.4) makes it inevitable that the amino acid composition of the seeds of individual plant species will differ markedly from one another. This is important when considering the suitability of a particular seed as a protein source for human nutrition. There is no seed which has the ideal balance of essential amino acids as far as human nutrition is concerned. Cereal seeds, for example, are deficient in lysine, which is present in low concentrations in prolamins and glutelins. Legume seeds are deficient in methionine as a result of the preponderance of globulin in their protein reserve. An interesting development in the field of plant breeding has been the production of new varieties of maize, 'opaque-2' and 'flowery-2' which possess increased ratios of lysine-rich albumins and globulins to lysine-deficient prolamins when compared with conventional varieties. A nutritionally more acceptable maize meal can be produced from these high-lysine varieties. (Unfortunately, these new varieties are not particularly high-yielding in terms of grain production.)

It must be appreciated that the terms we use to classify seed proteins (albumins, globulins, prolamins and glutelins) are broad ones and that each class includes proteins whose primary structure differs from plant to plant. Thus the globulins present in barley seeds will not be identical to the globulins present in soya bean and will even bear a different specific name. The prolamin from barley is referred to as 'hordein', that of wheat is known as 'gliadin' while maize prolamin is called 'zein'. 'Hordenin' is the name given to barley glutelin and 'glutenin' is the name for wheat glutelin. Both have a much higher lysine content than the glutelin of maize. A number of proteins of the same class may be present in the same seed (glutelins excepted).

For example, wheat prolamin (gliadin) is composed of separate proteins referred to as α-, β-, γ- and ω-gliadins, each with different molecular weights. The globulins may be similarly classified, although in some plants they are given individual names, for example, the globulins vicilin and legumin which occur together in many legume seeds. (To further bedevil an already confused nomenclature system, proteins may be classified according to their sedimentation coefficient when centrifuged – thus, vicilin is also known as 7 s globulin.) All seeds normally contain a number of different albumins as this class includes the numerous seed enzymes.

Table 8-2   Common names of some seed proteins.

Globulins	Prolamins	Glutelins
Vicilin (legumes)	Hordein (barley)	Glutenin (wheat)
Legumin (legumes)	Zein (maize)	Hordenin (barley)
Convicilin (pea)	Gliadin (wheat)	
	Kafirin (sorghum)	
	Avenin (oats)	

Many seed proteins are large and complex molecules. Vicilin has been shown by gel electrophoresis to be composed of three polypeptide molecules each having an average molecular weight of 58 000 daltons; legumin has a quaternary structure of twelve such units, each with an average molecular weight of 33 000 daltons. Glutelins have also been shown to consist of several polypeptide components, for example, wheat glutelin has a molecular weight of approximately 6 million daltons and comprises some 300 subunits.

8.1.2   Protein bodies

The reserve proteins of seeds have been shown to be localized in protein bodies or aleurins within the cells of cotyledons and endosperms. These are spherical or oval bodies surrounded by a single membrane and appear to have a different origin in cotyledons and endosperm (Fig. 8-1).

In cotyledons, the proteins (usually globulins) are formed by the ribosomes on the rough endoplasmic reticulum (ER) of the storage cells. The proteins are then injected into the cisternae of the ER which bud off vesicles that migrate to the Golgi bodies (dictyosomes), in which condensation of the protein takes place. Vesicles released from the maturing face of the Golgi bodies migrate to cell vacuoles, where the membranes of vesicles and vacuoles fuse, depositing the protein within the vacuoles. On filling with protein, these become the protein bodies.

In the endosperm, which usually stores prolamins, the process appears to be more direct, the dictyosome-produced vesicles or the endoplasmic reticulum itself giving rise to the protein bodies which have been shown to contain inclusions within their proteinaceous matrix. These are of two main types – crystalloid (probably crystallized prolamin) and globoid (consisting of phytin, a salt of inositol phosphate). A fair amount of controversy has developed over the nomenclature of protein bodies, centred on the presence or absence of these inclusions.

In cereal seeds the cells of the outermost layer of the endosperm, called the aleurone layer, are particularly rich in protein bodies. These protein bodies are referred to by many workers as 'aleurone grains', a name which at one time was given to all protein bodies. In the manufacture of white flour from wheat this aleurone layer unfortunately remains with the bran. Thus 'white' bread has a much lower protein content than bread baked from wholemeal flour. (Note that protein crystals also occur outside protein bodies, for example, in leucoplasts of the root tip of bean plants.)

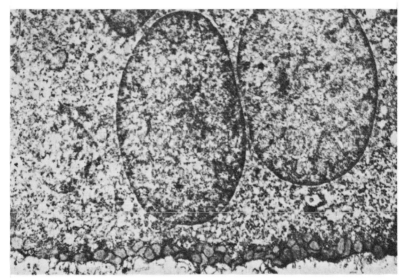

Fig. 8-1 Electron micrograph of protein bodies in a cotyledonary cell of *Pisum sativum* (× 24 640). (Mollenhauer, H.H. In Beevers, L. (1976). *Nitrogen metabolism in plants*. Edward Arnold, London.)

8.2 Seed germination

The onset of seed germination is precipitated by a combination of favourable environmental and internal factors which raises the rate of a number of previously dormant metabolic processes in the tissues of the seed. An early effect is the beginning of mobilization of the protein reserves of the seed, which takes place in the protein bodies. There is some controversy as to whether proteolytic enzymes (proteases) are present in the protein bodies in an inactive form prior to the commencement of

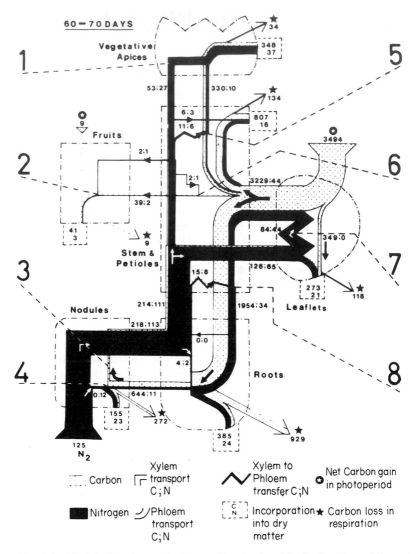

Fig. 9-1   Model of carbon and nitrogen flow in the white lupin. (Pate, J.S. *et al.* (1979). *Plant Physiol.*, **63**, 730–38.)

nitrogenous compounds, 4mg being used in the root while the reminder (214mg) remains in the xylem stream for delivery to the shoot.

   In the case of nitrogen, the only source for the plant has been the fixation of gaseous N_2 by the nodules (125mg per 10 days). The bulk of the N (111mg) has been exported from root to shoot, with 65mg per 10 days arriving in the leaves. 21mg N of this have been deposited in the leaf, 44mg being re-exported, mainly to the root (34mg). In the root, 24mg have been incorporated into root tissue and 11mg exported to the nodules over the 10 day period. It is interesting to note that 90 per cent of the nitrogen acquired by the root over this time has been from the shoot and not as a direct import from the N-rich xylem stream leaving the nodules. The fruits have received 3mg N over the 10 days, 2mg arriving via the phloem and 1mg via the xylem.

Further information regarding the flux of C and N round the plant may be obtained from the other details provided in the figure.

Predictions which can be made from this model are listed below (1–8) with the site of the application of each prediction being located by the corresponding large number in the figure.

(1) Vegetative apices received 73 per cent of their N and 14 per cent of their C through the xylem.

(2) Fruits received 38 per cent of their N and 6 per cent of their C through the xylem.

(3) 34 per cent of the C supplied to the nodule returned to the shoot attached to fixation products.

(4) 48 per cent of the N incorporated into growing nodules was supplied from the shoot in the phloem.

(5) Xylem to phloem transfer in the stem provided vegetative apices with 60 per cent of the N they received through phloem.

(6) Xylem to phloem transfer in the stem provided fruits with 70 per cent of the N they received through phloem.

(7) 68 per cent of the N received by leaves via the xylem was transferred to phloem and exported from leaves with photosynthate.

(8) Xylem to phloem transfer in the stem provided the nodulated root with 23 per cent of the N it received through phloem.

Models of this type are of great value in comprehending the quantity and complexities of nitrogen exchange within the whole plant, and in providing data from which to predict the effects of various treatments or environmental conditions on plant productivity.

10 Nitrogen and Man

In the present-day world, Man's concern with nitrogen is focused primarily on two major fields in which the element plays a prominent role: ecology and agriculture.

10.1 Nitrogen in ecology

One of Man's main objectives in the study of ecology is to understand the workings of the world's major ecosystems in order to preserve them, and the global environment of which they form a part, in the face of his many environmentally-damaging activities. This ideal is not altruistic for were he to allow the destruction of his natural environment, Man must inevitably precipitate his own extinction.

In nearly all ecosystems, nutrient-rich or nutrient-poor, nitrogen is the element which most often limits the growth of plants. Thus, whether one is investigating Tundra, Mediterranean, Desert, Deciduous Forest, Coniferous Forest, Grassland, or Savanna ecosystems, a prime element of the research programme for that system is always an examination of its nitrogen component. This involves a study of the following ecosystem properties and processes:

(*a*) combined nitrogen input from wet and dry deposition and from fixation brought about by soil and symbiotic organisms;

(*b*) losses of combined nitrogen from run-off and leaching, combustion, denitrification and ammonia volatilization;

(*c*) estimates of soil nitrogen content in its different forms, each of which is variably available for plant uptake (organic nitrogen, nitrate, bound ammonium, exchangeable ammonium);

(*d*) estimates of the nitrogen content of above- and below-ground biomass and necromass;

(*e*) estimates of litter and the rate at which mineralization and nitrification of the nitrogen fraction of litter takes place.

From this information a model of the cycling of nitrogen through the ecosystem can be constructed and used to predict the likely effects of various environmental stresses and management regimes on this process.

Naturally, studies of this nature are never made in isolation, but are carried out in conjunction with the investigation of other environmental factors which are likely to affect the ecosystem (e.g. energy flow, water relations, and the cycling of phosphorus and sulphur).

Studies such as these have helped to highlight unwise ecosystem management practices, for example the conversion of those apparently superfertile regions of the earth's surface, capable of sustaining the luxuriant growth of rain forest, into crop production. Many of these regions are actually nutrient-poor, but possess a tightly closed nitrogen cycle within the forest ecosystem, with much of the combined nitrogen locked into the plants themselves. Once the forest cover is removed, the nitrogen-poor soils are unable to sustain crop growth for more than one season, and many years must elapse before the environmental damage can be repaired by forest regeneration.

10.2 Nitrogen in agriculture

Protein production for the feeding of the world's human population is one of the main preoccupations of agriculture. In spite of the tremendous strides made in the application of scientific knowledge to improve agricultural productivity in recent years, mankind's number one killer disease remains, not cancer nor heart ailments, but starvation, that is, a lack of dietary protein and calories. In terms of protein supply this fact seems somewhat surprising in view of apparent overproduction by global agriculture. N.W. Pirie made the following calculations in 1973:

World population	3 600 000 000
Daily world protein requirement (based on the FAO figure of 50 g protein per 60 kg person per day)	180 000 000 kg
Estimated daily world protein production (plant crops)	460 000 000 kg

Nevertheless, recent statistics have shown that about one-third of the population of the underdeveloped nations (i.e. about 20 per cent of the world's population, or 800 million people), are malnourished, with some 60 000 000 actually on the starvation borderline.

The reasons for this unhappy situation are mainly the following.

(a) *Spoilage of food material*   Owing to poor storage and transport facilities, particularly in the underdeveloped countries, nealy 35 per cent of the world's harvests are lost to pests such as rats and weevils, or to spoilage by bacteria and fungi.

(b) *Meat production*   Because of the preference shown by human beings for meat products, over 60 million tonnes per annum of protein-rich oilseed (half the world's supply) is used in feeding animals destined to provide meat for human consumption. Animals are notoriously inefficient concentrators of plant proteins; it takes 5 kg of plant protein to produce 1 kg of animal protein, an expensive method of concentrating protein!

(c) *Uneven world distribution of protein production*   In the developed regions of the world, there is an overproduction of food materials and one frequently hears of the butter mountains and milk lakes which the European Economic Community has difficulty in consuming. In the underdeveloped regions of the world there is, on the other hand, gross underproduction of foodstuffs in spite of the fact that most of these areas lie within or close to the tropics, and therefore have the potential for growing two crops per year; such regions convert only about 5×10^{-2}per cent of the solar energy they receive into crop plants, whereas in parts of Europe this figure can be as high as 33×10^{-2}per cent. The reasons for this disparity in productivity are varied, complex and, as yet, imperfectly understood. An obvious remedy for this geographical imbalance of protein/energy production would be to transfer the surplus from regions of overproduction to those of underproduction. Unfortunately this simple answer is impractical because of the tremendous costs involved in the transport and final distribution of the food material, and the difficulties involved in the administration of such a colossal scheme.

Has the science of agriculture been able to ameliorate significantly the miseries of malnutrition in the modern world? The answer to this question can only be a resounding affirmative when one appreciates the fact that between the years 1950 and 1975 the 'Green Revolution' has been able to almost double the world's food supply. Its impact has regrettably been diminished by an exploding world population which has grown at almost the same rate; were it not for this agricultural advance the dimensions of hunger and malnutrition would be far greater than they are now.

This Green Revolution has been achieved mainly by the following means:

(*a*) The breeding of new hybrid strains of crop plants with greatly enhanced grain production under suitable fertilizer regimes, improved essential amino acid content of the grain protein, superior physical properties (e.g. dwarf varieties of wheat and rice which are less susceptible to climate damage and 'lodging' than their progenitors), and improved resistance to fungal disease and environmental stress (e.g. water shortage).

(*b*) The introduction of energy-intensive mechanized techniques for soil preparation, sowing, fertilizing, spraying, reaping and product distribution.

(*c*) The development of environmentally-tolerable insecticides to combat the ravages of insects before and after harvesting.

We are all well aware of the fact that the earth's human population is currently undergoing an explosion which commenced in Europe at the time of the industrial revolution. Although the human population growth rate in the developed regions of the world is now approaching zero, the underdeveloped regions are still at their explosive stage of population growth and it is anticipated that the number of human beings present on this planet will increase from the current 4 800 000 000 (1984) to some 6 000 000 000 by the turn of the century. Can the intensity of the Green Revolution be increased to cater for the needs of this human avalanche? The answer is probably negative for the simple reason that the miracles of the Green Revolution have been achieved only by the application of large amounts of energy, nearly all of which has been derived from that dwindling and expensive resource, fossil fuel.

A consideration of only one aspect of the Green Revolution will serve to demonstrate the magnitude of energy expenditure involved. The recently developed high-yielding grain plants have increased food production prodigiously. However, plants cannot manufacture protein if they are not supplied with combined nitrogen; in order to realize the full yield potential of these 'wonder' plants vast quantities of nitrogen fertilizer have had to be supplied, produced mainly by the energy-demanding Haber-Bosch process. The escalation of the use of nitrogen fertilizer as a consequence of the Green Revolution and the cost thereof has been estimated as follows:

Annual world nitrogen fertilizer consumption
during 1950: 4 000 000 tons (value: $800 000 000)
during 1975: 40 000 000 tons (value: $10 000 000 000)

To expand the Green Revolution into the 21st Century using present techniques is likely to be prohibitively expensive, with a projected annual nitrogen fertilizer consumption of 200 000 000 tons (value: $200 000 000 000) for the year 2000. If even the present world population were to be nourished at the USA level (where the climax of the Green Revolution may be observed) by means of energy intensive agriculture, the annual energy cost is thought equal to 1 500 000 000 000 gallons of petrol and would completely exhaust the world's known oil reserves within 12 years.

There is thus an urgent necessity to produce food by less energy intensive methods than we are employing at the moment (as well as to check the world population expansion, a topic which is outside the scope of this book). With respect to the conservation of energy expensive nitrogen fertilizer, the following measures are being adopted or are under scientific investigation.

(a) The expansion of cultivation of legume crops, particularly soya bean, which produce protein-rich seeds and require no nitrogen fertilization. In the USA, vast areas formerly under maize cultivation are now being planted with soya bean and it is estimated that in that country alone the value of the nitrogen fixed by these plants exceeds $3 000 000 000 p.a. Much research is being conducted at the moment to establish varieties of legume crops which can be cultivated in the world's different climatic regions and whose products are acceptable to local populations (a very important factor). The conversion of animal pastures from grass to mixed clover – grass association is another important fertilizer nitrogen conservation strategy, and in Australia many millions of dollars have been saved in recent years by the introduction of this practice.

Crop rotation, a practice going back to the time of 'Farmer' George and 'Turnip' Townshend (George III of Britain and the Secretary of State to George I, respectively) in the 1700s, has also proved to be an effective method of soil nitrogen enrichment when one of the rotated crops is a legume.

(b) The increase in the efficiency of nitrogen fertilizer usage. We have already seen that much of the nitrogen fertilizer added to the soil by farmers is washed into rivers by rain or denitrified to the atmosphere. A great deal of research is being conducted at the moment to find methods of keeping the nitrogen in the soil. These methods include staggered addition of fertilizer to the soil to coincide with times of peak demand by the plant, the use of ammonium fertilizer, which is held firmly by the soil (in contrast to nitrate), and the addition of compounds such as nitrapyrin which prevent the oxidation of ammonium to nitrate in the soil. The growth-promoting effect of a mixed nitrate-ammonium source noted by Cox, Reisenauer and colleagues in a number of plants is another focus of research and, in conjunction with the use of nitrification inhibitors, promises enhanced efficiency of utilization of fertilizer nitrogen if soil ammonium-nitrate ratios can be controlled.

(c) The application of genetic engineering. This new science holds great potential for the improvement of the food-producing characters of crop plants. One need only contemplate the benefits to mankind of the transfer of the *nif* gene from *Rhizobium* to, say, *Zea mays* to realize the enormous impact this science could have on agriculture. Unfortunately, this particular transfer is a difficult one to achieve because of the complexity of the nitrogen-fixing process. Not only do the genes for the production of the sub-components of nitrogenase need to be transferred, but ancillary ones as well, such as those necessary for the production of leghaemoglobin or other mechanisms to ensure an anaerobic environment for the operation of the enzyme.

With these and other measures, it is anticipated that global agriculture will develop to the state where it can adequately nourish the whole of mankind – not at the present level of the most affluent nations of the world (the earth has the resources to feed about 1 000 000 000 people at this level, i.e. about one quarter of the world's present population!) but at a minimum level consistent with the maintenance of a healthy life. However more efficient agriculture may become, it is doubtful whether the resources of the earth will be able to sustain more than 20 000 000 000 people. There is little doubt, however, that most of their combined nitrogen will be supplied by the bacterial genus *Rhizobium*, or its *nif* genes!

Further Reading

ALEXANDER, M. (1977). *Introduction to Soil Microbiology* (2nd ed.). John Wiley & Sons, New York.

BARKER, G.R. (1982). *Chemistry of the Cell*. Studies in Biology no. 13 (2nd ed). Edward Arnold, London.

BEEVERS, H. (1977). *Nitrogen Metabolism in Plants*. Edward Arnold, London.

BELL, E.A. and CHARLWOOD, B.V. (eds)(1980). Secondary plant products. *Encyclopaedia of Plant Physiology (NS) Vol. 8*. Springer-Verlag, Berlin.

BOULTER, D. and PARTHIER, B. (eds)(1982). Nucleic acids and proteins in plants. I, structure, biochemistry and physiology of proteins. *Encyclopaedia of Plant Physiology (NS) Vol. 14A*. Springer-Verlag, Berlin.

BRAY, C.M. (1983). *Nitrogen Metabolism in Plants*. Longman, London.

CLARKE, F.E. and ROSSWALL, T. (eds)(1981). Terrestrial nitrogen cycles. *Ecological Bulletins No. 33*. Swedish Natural Sciences Research Council, Stockholm.

FOWDEN, L., LEA, P.J. and BELL, E.A. (1979). The non-protein amino acids of plants. *Advances in Enzymology*, **50**, 117-175.

GUERRERO, M.G., VEGA, J.M. and LOSADA, M. (1981). The assimilatory nitrate-reducing system and its regulation. *Ann. Rev. Plant Physiol.*, **32**, 169-204.

HAYNES, R.J. and GOH, K.M. (1978). Ammonium and nitrate nutrition of plants. *Biol. Rev.*, **53**, 465-510.

HEWITT, E.J. (1975). Assimilatory nitrate-nitrite reduction. *Ann. Rev. Plant Physiol.*, **26**, 73-158.

LÄUCHLI, A. and BIELESKI, R.L. (eds)(1983). Inorganic nitrogen nutrition. *Encyclopaedia of Plant Physiology (N.S.) Vol. 15A*. pp. 241-369. Springer-Verlag, Berlin.

LEA, J.A. and STEWART, G.R. (1978). Ecological aspects of nitrogen assimilation. *Advances in Botanical Research*, **6**, 1-43.

LIKENS, G.E. (1981). Some perspectives of the major biogeochemical cycles. *Scope 17*. John Wiley & Sons, New York.

MIFLIN, B.J. (ed.)(1980). Amino acids and derivatives. *The Biochemistry of Plants Vol. 5*. Academic Press, New York.

MIFLIN, B.J. and LEA, P.J. (1977). Amino acid metabolism. *Ann. Rev. Plant Physiol.* **28**, 299-329.

MIFLIN, B.J. and LEA, P.J. (1982). Ammonia assimilation and amino acid metabolism. In BOULTER, D. and PARTHIER, B. (eds). Nucleic acids and proteins in plants I. *Encyclopaedia of Plant Physiology (NS) Vol. 14A*. Springer-Verlag, Berlin.

MARCUS, A. (ed)(1981). Proteins and nucleic acids. *The Biochemistry of*

Plants Vol. 6. Academic Press, New York.

PATE, J.S. (1973). Uptake, assimilation and transport of nitrogen compounds by plants. *Soil Biol. Biochem.* **5**, 109–119.

PATE J.S. (1975). Exchanges of solutes between phloem and xylem, and circulation in the whole plant. In ZIMMERMAN, M.H. and MILBURN, J.A. (eds). *Encyclopaedia of Plant Physiology (N.S.) Vol. 1.* Springer-Verlag, Berlin.

PATE, J.S. (1980). Transport and partitioning of nitrogenous solutes. *Ann. Rev. Plant Physiol.*, **31**, 313–40.

POSTGATE, J. (1978). *Nitrogen Fixation.* Studies in Biology no. 92. Edward Arnold, London.

WYNN, C.H. (1979). *The Structure and Function of Enzymes.* Studies in Biology no. 42 (2nd ed.). Edward Arnold, London.

References

(Additional to those appearing in tables and figures)

BEEVERS, L. (1976). *Nitrogen Metabolism in Plants*. Edward Arnold, London.

BENZIONI, A., VAARDA, Y. and LIPS, S.H. (1970). Correlations between nitrate reduction, protein synthesis and malate accumulation. *Physiol. Plant.* **23**, 1039–1047.

DELWICH, C.C. and LIKENS, G.E. (1982). In Some Perspectives of the Major Geochemical Cycles. *Scope 17*, 43. Wiley Ambers and Sons, London.

EVANS, H.J. and NASON, A. (1953). Pyridine nucleotide-nitrate reductase from extracts of higher plants. *Plant Physiol.* **28**, 233–254.

HAGEMAN, R.H., CRESSWELL, C.F. and HEWITT, E.J. (1962). Reduction of nitrate, nitrite and hydroxylamine to ammonia by enzymes extracted from higher plants. *Nature* **193**, 247–250.

KRAJINA, V.B., MADOC-JONES, S. and MELLOR, G. (1973). Ammonium and nitrate economy of some conifers growing in Douglas-fir communities of the Pacific northwest of America. *Soil Biol. Biochem.* **5**, 143–147.

LEA, P.J. and MIFLIN, B.J. (1974). An alternative route for nitrogen assimilation in higher plants. *Nature* **251**, 614–616.

LEWIS, O.A.M., JAMES, D.M. and HEWITT, E.J. (1982). Nitrogen assimilation in barley (*Hordeum vulgare* L. cv. Mazurka) in response to nitrate and ammonium nutrition. *Ann. Bot.* **49**, 39–49.

LEWIS, O.A.M. and PATE, J.S. (1973). Significance of transpirationally derived nitrogen in protein synthesis in fruiting plants of pea (*Pisum sativum* L.). *J. Exp. Bot.* **24**, 596–606.

MIFLIN, B.J. and LEA, P.J. (1976). The pathway of ammonia assimilation in plants. *Phytochemistry* **15**, 873–885.

PATE, J.S., LAYZELL, D.B. and McNEIL, D.L. (1979). Modelling the transport and utilization of carbon and nitrogen in a nodulated legume. *Plant Physiol.* **63**, 730–738.

PATE, J.S. (1973). Uptake, assimilation and transport of nitrogen compounds by plants. *Soil Biol. Biochem.* **5**, 109–119.

PIRIE, N.W. (1974). World Food Supplies. In *Medicine* **28**, 76–81.

SIMS, A.P. and FOLKES, B.F. (1964). A kinetic study of the assimilation of (^{15}N) ammonia and the synthesis of amino acids in an exponentially growing culture of *Candida utilis*. *Proc. Roy. Soc. B.* **159**, 479–502.

SOARES, M.I.M., LIPS, S.H. and CRESSWELL, F.C. (1985). *Physiol. Plant.* (in press).

TEMPEST, D.W., MEERS, J.L. and BROWN, C.M. (1979). Synthesis of glutamate in *Aerobacter aerogenes* by a hitherto unknown route. *Biochem J.* **117**, 405–407.

Index